Elevation Models for Geoscience

Geological Society books refereeing procedures

The Society makes every effort to ensure that the scientific and production quality of its books matches that of its journals. Since 1997, all book proposals have been refereed by specialist reviewers as well as by the Society's Books Editorial Committee. If the referees identify weaknesses in the proposal, these must be addressed before the proposal is accepted.

Once the book is accepted, the Society Book Editors ensure that the volume editors follow strict guidelines on refereeing and quality control. We insist that individual papers can only be accepted after satisfactory review by two independent referees. The questions on the review forms are similar to those for *Journal of the Geological Society.* The referees' forms and comments must be available to the Society's Book Editors on request.

Although many of the books result from meetings, the editors are expected to commission papers that were not presented at the meeting to ensure that the book provides a balanced coverage of the subject. Being accepted for presentation at the meeting does not guarantee inclusion in the book.

More information about submitting a proposal and producing a book for the Society can be found on its web site: www.geolsoc.org.uk.

It is recommended that reference to all or part of this book should be made in one of the following ways:

Fleming, C., Marsh, S. H. & Giles, J. R. A. (eds) 2010. *Elevation Models for Geoscience.* Geological Society, London, Special Publications, **345**.

Giles, J. R. A., Marsh, S. H. & Napier, B. 2010. Dataset acquisition to support geoscience. *In*: Fleming, C., Marsh, S. H. & Giles, J. R. A. (eds) *Elevation Models for Geoscience.* Geological Society, London, Special Publications, **345**, 135–143.

GEOLOGICAL SOCIETY SPECIAL PUBLICATION NO. 345

Elevation Models for Geoscience

EDITED BY

C. FLEMING, S. H. MARSH and J. R. A. GILES
British Geological Survey, Keyworth, UK

2010
Published by
The Geological Society
London

THE GEOLOGICAL SOCIETY

The Geological Society of London (GSL) was founded in 1807. It is the oldest national geological society in the world and the largest in Europe. It was incorporated under Royal Charter in 1825 and is Registered Charity 210161.

The Society is the UK national learned and professional society for geology with a worldwide Fellowship (FGS) of over 10 000. The Society has the power to confer Chartered status on suitably qualified Fellows, and about 2000 of the Fellowship carry the title (CGeol). Chartered Geologists may also obtain the equivalent European title, European Geologist (EurGeol). One fifth of the Society's fellowship resides outside the UK. To find out more about the Society, log on to www.geolsoc.org.uk.

The Geological Society Publishing House (Bath, UK) produces the Society's international journals and books, and acts as European distributor for selected publications of the American Association of Petroleum Geologists (AAPG), the Indonesian Petroleum Association (IPA), the Geological Society of America (GSA), the Society for Sedimentary Geology (SEPM) and the Geologists' Association (GA). Joint marketing agreements ensure that GSL Fellows may purchase these societies' publications at a discount. The Society's online bookshop (accessible from www.geolsoc.org.uk) offers secure book purchasing with your credit or debit card.

To find out about joining the Society and benefiting from substantial discounts on publications of GSL and other societies worldwide, consult www.geolsoc.org.uk, or contact the Fellowship Department at: The Geological Society, Burlington House, Piccadilly, London W1J 0BG: Tel. +44 (0)20 7434 9944; Fax +44 (0)20 7439 8975; E-mail: enquiries@geolsoc.org.uk.

For information about the Society's meetings, consult *Events* on www.geolsoc.org.uk. To find out more about the Society's Corporate Affiliates Scheme, write to enquiries@geolsoc.org.uk.

Published by The Geological Society from:
The Geological Society Publishing House, Unit 7, Brassmill Enterprise Centre, Brassmill Lane, Bath BA1 3JN, UK

(*Orders*:   Tel. +44 (0)1225 445046, Fax +44 (0)1225 442836)
Online bookshop: www.geolsoc.org.uk/bookshop

The publishers make no representation, express or implied, with regard to the accuracy of the information contained in this book and cannot accept any legal responsibility for any errors or omissions that may be made.

British Library Cataloguing in Publication Data

A catalogue record for this book is available from the British Library.
ISBN 978-1-86239-313

Typeset by Techset Composition Ltd, Salisbury, UK
Printed by MPG Books Ltd, Bodmin, UK

Distributors

North America
For trade and institutional orders:
The Geological Society, c/o AIDC, 82 Winter Sport Lane, Williston, VT 05495, USA
Orders: Tel. +1 800-972-9892
         Fax +1 802-864-7626
         E-mail: gsl.orders@aidcvt.com

For individual and corporate orders:
AAPG Bookstore, PO Box 979, Tulsa, OK 74101-0979, USA
Orders: Tel. +1 918-584-2555
         Fax +1 918-560-2652
         E-mail: bookstore@aapg.org
         Website: http://bookstore.aapg.org

India
Affiliated East-West Press Private Ltd, Marketing Division, G-1/16 Ansari Road, Darya Ganj, New Delhi 110 002, India
Orders: Tel. +91 11 2327-9113/2326-4180
         Fax +91 11 2326-0538
         E-mail: affiliat@vsnl.com

Contents

Introducing elevation models for geoscience

C. FLEMING*, S. H. MARSH & J. R. A. GILES

British Geological Survey, Kingsley Dunham Centre, Keyworth, Nottingham NG12 5GG, UK

Corresponding author (e-mail: ccot@bgs.ac.uk)

Elevation data are a critical element in any geoscience application. From the fundamentals of geological mapping to more advanced three-dimensional (3D) modelling of Earth systems there must be an understanding of the shape of the Earth's surface. Vast amounts of digital elevation data exist, from large-scale global datasets to smaller-scale regional datasets, and in many cases datasets have been merged to improve scale and accuracy. For each application decisions must be made on which elevation data are appropriate. This will depend on many factors including the cost, resolution and accuracy of the data.

The types of data discussed in this special publication include: ASTER (Advanced Spaceborne Thermal Emission and Reflection Radiometer), LiDAR (Light Detection And Ranging) – terrestrial and airborne, NEXTMap, SRTM (Shuttle Radar Topography Mission) and multibeam bathymetry. Applications covered include: landslide mapping, coastal erosion, glacial deposits and hazard mapping, and some of the issues discussed include: accuracy analysis, derived product creation, software comparisons and copyright considerations (Table 1). Since some of the papers were written for the Special Publication certain datasets have evolved and been created; for example, the GDEM global elevation dataset derived from ASTER data. This illustrates the fast moving nature of this field.

With the proliferation in data available for the production of digital elevation models (DEMs) it is increasingly important to understand how to use the raw data correctly and effectively. **Giglierano** discusses the use of LiDAR for natural resource mapping applications, and states how a 'black box' approach is dangerous and that knowledge of the data being used is essential, especially as more non-specialists begin to use the data. Many users reduce the resolution of the DEM to shorten processing time and also decrease the amount of space required to store the data. This obviously reduces the detail in the data and removes some of the more subtle ground features. Products derived from inaccurate raw data will have inherent inaccuracies of their own.

NEXTMap data have several derived products and it is important to understand how they have been derived, either directly from the raw data or from a model created to strip features to the ground surface. Natural and man-made features on the ground surface can interfere with the acquisition of a perfect representation of the terrain. Buildings and other regular structures can be removed relatively easily due to their recognizable angular shape and, in most cases, known height. Trees and other natural features are not so easy to remove. In some datasets, for example LiDAR, the spacing of the data points is small enough that some of the data will penetrate the vegetation and provide a measurement of the ground surface. Even if some interpolation is required, it is still possible to develop a model of the ground surface beneath vegetation based on a few data points rather than none at all. The spacing and wavelength of the NEXTMap data is much greater than that of LiDAR and so very few radar pulses penetrate to the ground surface. All of the data beneath vegetation must, therefore, be modelled and so this inevitably leads to inaccuracies in the 'bare earth model' created.

Issues of data inaccuracies in bare earth models have been described in detail in **Rutter et al.**, especially their use in hydrogeological modelling. Hydrogeological modelling is very dependent on an accurate DEM. Where ground surfaces have been modelled rather than measured, incorrect flow of water can occur; for example, rivers flowing upstream. **Rutter et al.** describe how it is important to select the correct DEM for the specific application and discuss the use of derived geological datasets on a national scale. Many UK datasets are based on NEXTMap data. The main issue highlighted in this paper is that of accuracy in the NEXTMap DTM, which is a model of the land surface rather than the measured surface that includes buildings and trees. The development of an uncertainty layer is suggested so that the user can decide whether a layer is accurate enough for their purposes.

The amount of relief present in an area can also dictate how accurate the DEM for that area will be. Recorded spot heights measured from ASTER DEMs, for example, are more accurate in lower relief areas (**Nightingale & Morgan** and **Cziferszky et al.**). Conversely, contours derived from such a low-relief dataset can be less accurate owing to the lack of features present from which to derive contours. Shadows, distortion of higher

From: FLEMING, C., MARSH, S. H. & GILES, J. R. A. (eds) *Elevation Models for Geoscience.*
Geological Society, London, Special Publications, **345**, 1–4.
DOI: 10.1144/SP345.1   0305-8719/10/$15.00 © The Geological Society of London 2010.

Table 1. *Summary of applications and sensors discussed in this Special Publication*

Type of sensor	Application						
	Accuracy analysis	Geological and natural resource mapping	Hazards	Derived products	Environmental	Glaciation	Software comparisons
SRTM	Smith Cziferszky *et al.*	Crippen	Crippen	Smith	Nightingale & Morgan		Cziferszky *et al.*
ASTER	Danneels *et al.* Smith Hall & Tragheim Cziferszky *et al.*	Cziferszky *et al.*	Nightingale & Morgan	Smith Danneels *et al.* Giles *et al.*	Nightingale & Morgan	Cziferszky *et al.*	Cziferszky *et al.* Danneels *et al.* Nightingale & Morgan
NEXTMap	Hall & Tragheim Smith	Hall *et al.* Giles *et al.*	Rutter *et al.* Mackenzie *et al.*	Smith Giles *et al.*	Rutter *et al.* Mackenzie *et al.*	Hall *et al.*	
Bathymetry		Tappin	Tappin		Tappin		
LiDAR	Smith Giglierano	Giglierano		Smith Giglierano Giles *et al.*			
Terrestrial LiDAR			Hobbs *et al.*		Hobbs *et al.*		

slopes and the size of the pixels all influence how accurate the DEM will be. Thus, in lower relief areas these factors will not be as much of an issue as in higher relief areas.

In order to improve the accuracy of the ASTER DEM **Nightingale & Morgan** combined the co-registered ASTER and SRTM DEM using manually entered spot heights from the SRTM at exact pixel locations within both models. Some loss of detail may have occurred owing to the difference in resolution between the two datasets but the improvement in the accuracy of the height data from the resulting ground control point (GCP)-corrected, fused ASTER–SRTM DEM is significant. The study area chosen to test this refinement is an area of high relief in China where accurate data are required for detailed monitoring of soil erosion and slope stability upstream of the Three Gorges Dam.

The accuracy of ASTER DEMs is also discussed by **Hall & Tragheim** in comparison with NEXTMap data. **Hall & Tragheim** use a study area in South Wales, where both ASTER and NEXTMap data are available, to investigate differences in the two DEMs so that possible regular inaccuracies within the ASTER DEM can be highlighted and then identified in areas where no coincident/comparable data are available. Profile and contour comparisons have been made to allow any variation in accuracy to be observed. It was calculated that 95% of the ASTER DEM points were within ± 20 m of the NEXTMap DEM. This would allow 40 m contours to be generated from overseas ASTER data with a greater level of confidence.

Improvements in the method of satellite acquisition of digital elevation data have advanced this field markedly. **Cziferszky et al.**, **Hall & Tragheim** and **Nightingale & Morgan** all comment on the change from across-track data acquisition (SPOT1–4) to along-track (ASTER, SRTM). The advantage is the coincident data as the stereo pair is acquired in the same light, weather and ground conditions. This is a clear step forward in the production of good quality DEMs.

Cziferszky et al. suggest that ASTER is the most useful elevation dataset available for the polar regions, where there are no SRTM data and airborne data are difficult to acquire. **Cziferszky et al.** provide an assessment of ASTER elevation data over mountainous, glaciated terrain, in Antarctica. This challenging environment creates several issues when trying to use elevation models from satellite platforms. Different terrain types present their own challenges and this paper discusses how best to generate DEMs from ASTER data, achieving the lowest errors possible, over the different terrain types present. Again, the accuracy of ASTER DEMs is investigated in this

paper with airborne photogrammetric data used as a comparison and three different software packages assessed. It is concluded that the accuracy of the DEM is linked to the steepness of the terrain, with higher relief yielding much reduced accuracy.

Advances in software have also made the production of accurate DEMs possible. Several of the papers in this volume investigate the differences in the production methods of the software used to create DEMs. **Cziferszky et al.** investigate the differences in accuracy between three different software packages, whereas **Nightingale & Morgan** and **Danneels et al.** describe how they have modified the raw data to improve DEM accuracy using their own scripts.

Nightingale & Morgan present a 'simple but effective' method to increase the level of accuracy of ASTER DEMs. SRTM data have been used to give better vertical accuracy, especially in high relief areas. This does come at a cost, however, as some detail found in the original ASTER data is inevitably lost following the merge process with the 90 m SRTM data. **Danneels et al.** present research into the filtering of a very noisy DEM created from a pair of ASTER images. Random, high-amplitude, sinks and mounds are present as well as a more regular east–west pattern. The filtering technique is designed to preserve the integrity of the data, and effectively removes noise and artefacts whilst no smoothing is applied. A technique known as mathematical morphology is used to define image pixel values.

Crippen provides the keynote paper, and talks about global topographical exploration and analysis with the SRTM and ASTER elevation models. This paper encapsulates the importance of topographical data. It discusses the use of ASTER and SRTM near-global digital elevation data, and its application to not only the Earth sciences but other disciplines such as climatology, hydrology and ecology. **Crippen** goes on to describe the method of acquisition, resolution, coverage and availability of both SRTM and ASTER data. Discussion then moves to the solution of certain limitations of each dataset. These limitations do not occur in the same places for each elevation model, and so often where one dataset is weaker the other is stronger and so 'holes' can be filled. For example, where ASTER is affected by clouds SRTM is not. Interpolation and rubber sheeting is used following the calculation of a difference image.

Texture is an important element when viewing DEMs, especially where subtle features need to be highlighted. Shaded relief models are often employed to highlight subtle surface features related to geology, etc., using low sun angles in the optimum orientation. Draping airborne or satellite imagery over an elevation model also enables more subtle features to be recognized. Using several datasets together with digital terrain data creates a much more powerful tool for the geoscientist. This is most evident in the 3D visualization of geological data (**Hall** *et al.*) but is equally useful for groundwater flooding prediction (**Mackenzie** *et al.*), interpretation of seabed morphology (**Tappin**) and the modelling of coastal change (**Hobbs** *et al.*), as discussed in this Special Publication.

Hall *et al.* describe how the British Geological Survey mapping teams have used NEXTMap Britain to enhance the geological mapping techniques employed. The data are used prior to mapping in the laboratory on screen, during mapping in paper form and on tablet PCs, and then after field mapping for refinement and the creation of 3D geological line work. The data, especially viewed as shaded relief models, have been particularly useful in the interpretation of glacial and proglacial deposits in the Vale of York. The automatic generation of landform features has also been made possible using NEXTMap data in areas where the bedrock is well featured.

McKenzie *et al.* use elevation models to predict areas at risk of groundwater flooding. They describe a methodology whereby groundwater flooding maps have been produced from the combination of geological and hydrogeological data with a DEM. National water-level data are inadequate/insufficient and so a water-level surface has been interpolated using known interactions between groundwater, surface water and DTM data. The maps created have been used to indicate areas susceptible to flooding owing to shallow groundwater levels.

Tappin introduces the use of multibeam bathymetry for the interpretation of seabed morphology. Three-dimensional visualization and complete seabed coverage enables the mapping of marine geohazards; this is equivalent to mapping undertaken on land. Three DEMs have been investigated in this paper in order to study submarine seabed failure related to tsunami events. The DEMs created from the multibeam bathymetry are used as the foundation for newly developed techniques to model seabed landslides that may have generated tsunami events.

Hobbs *et al.* discuss the use of terrestrial LiDAR to map and model coastal change. Mobile terrestrial LiDAR has been used to survey and monitor changes to the English coastline and estuarine environment. This technique is now used almost routinely in various applications, and is widely used in geoscience for hazard monitoring, quarry surveying and coastal erosion monitoring. Most of the work described in this paper involves the 'monitoring of active landslides on eroding coastlines' (**Hobbs** *et al.*). The set-up, positioning and accurate determination of the location of the scanner are discussed, because achieving a 'fix' on the location of a survey is difficult in the coastal zone owing to its dynamic nature and the lack of fixed reference points. Three-dimensional laser scan models can then be enhanced texturally and visually using digital photography, and used to interpret lithology, structure and geomorphology. All of this work is of great importance as climate change affects the rate and volume of coastal erosion, especially in the south of England.

Smith introduces the issue of copyright, especially in products derived from raw DEM data, within UK datasets. **Smith** uses his paper to highlight the issues faced by researchers when using UK elevation datasets. So much data are available that the first issue faced by the researcher is a choice as to which dataset is most suitable for their particular application. Intellectual property rights (IPR) and copyright issues are also discussed, with an emphasis on derived data products, how the data are accessed and the re-use of the data. Publication of research results may be affected by copyright and IPR. Even where the research is to be published non-commercially, licences may limit the graphical display of the elevation data. Any derived data, especially data products derived from several sources, are possibly subject to IPR restrictions from each vendor involved, which obviously affects publication and the sharing of output data.

Giles *et al.* outline the changing use of digital datasets from survey specific 'in-house' data to national datasets with a community of users worldwide. They also highlight how important elevation data are to the geologist, especially in the UK where the lack of rock exposure means that the technique known as feature mapping needs to be employed, whereby subtle breaks in slope are used to indicate possible changes in lithology. They also highlight some of the issues in managing and delivering elevation datasets, especially where national coverage is required.

As with all digital data, the product created is only as good as the raw data from which it is derived. Much is also dependent on the way in which the data are preprocessed. Any smoothing, interpolation or modelling that may lead to a loss of detail in the raw data may lead to a degradation in the accuracy of the DEM produced. Knowledge of any preprocessing is therefore essential if a true understanding of any subsequent model is to be achieved.

Global topographical exploration and analysis with the SRTM and ASTER elevation models

ROBERT E. CRIPPEN

NASA Jet Propulsion Laboratory, California Institute of Technology, 4800 Oak Grove Drive, Pasadena, CA 91109, USA (e-mail: Robert.E.Crippen@jpl.nasa.gov)

Abstract: One of the most fundamental geophysical measurements of the Earth is that which describes the shape of its land surface. Topographical data are required by virtually all Earth science disciplines engaged in studies at or near the land surface. Topography is also civilization's most heavily used non-atmospheric geophysical measurement. NASA's Shuttle Radar Topography Mission (SRTM) and ASTER (Advanced Spaceborne Thermal Emission and Reflection Radiometer) projects have each completed independent near-global digital elevation measurements at comparable resolutions that approach 30 m spatially and 10 m vertically. Exploration of these datasets provides a new perspective of our planet. Fusion of these datasets will produce a more complete global elevation database, and differentiation of these datasets can be used to quantify select geomorphic processes.

The shape of the Earth's surface is a dominant controlling factor in virtually every natural process that occurs there. It also significantly controls processes within the overlying atmosphere and indicates processes within the underlying lithosphere. Consequently, topographical information is important across the full spectrum of Earth sciences, including climatology, ecology, hydrology, glaciology and geology.

Precipitation, runoff, soil moisture, incident sunlight and temperature all vary with topography. Consequently, topography dominantly controls the local and regional distribution and character of vegetation. Erosion and sedimentation, and consequently soil formation and nutrient transport, are also strongly controlled by topography, and are important factors in ecological studies.

Topography strongly influences the location and magnitude of surface and subsurface water flux. The modelling of water supply and flood potential requires knowledge of the area's drainage extent, its slopes and the pattern of the drainage network. In many areas snowmelt is the major contributor to water supply, and the modelling of melt rates depends on knowledge of the radiation balance that is largely controlled by elevation, topographical shadowing and reflectance from neighbouring terrain.

Particularly in rugged terrain, topography is commonly the dominant variable in remote sensing imagery. Topographical shading affects the radiance measured at every wavelength and is consequently the statistical first principal component of many remotely sensed datasets. Meanwhile, atmospheric optical thickness varies inversely (and non-linearly) with topographical height, so that topography is an important factor in the atmospheric correction of remotely sensed data. Topography also distorts (with view angle) the geographic pattern recorded. In short, high-resolution and high-quality elevation data are essential in fully distinguishing terrain reflectance from terrain illumination and atmospheric optics, as well as in mapping the reflectance pattern with high spatial fidelity.

While topography controls many natural processes at and near the Earth's surface, many natural processes conversely control the topography. Consequently, to various degrees, topography records and reveals evidence of current and past natural processes. An obvious example is the development and occurrence of erosional and depositional fluvial landforms. However, tectonic, volcanic, glacial and gravitational processes also produce characteristic landforms that reveal past, ongoing and even potential change. The present is the key to the past (and future), and the past is the key to the present (and future). For example, topographical analysis is one of the primary means of determining the *current* global fault pattern, created by *past* and current processes, for assessing *future* seismic threats. Tectonic landforms, including surficial faults (commonly obvious as disruptions in the fluvial pattern), can indicate zones of earthquake hazards. Satellite imagery has greatly facilitated the mapping of the global tectonic pattern, revealed primarily in topographical shading, but topographical data facilitate more versatile and powerful means of landform analysis, not convolved with obscuring land cover patterns and not limited to analysis of shade patterns on a given day and time.

Topographical data also facilitate Earth surface visualization, a powerful tool that uniquely

From: FLEMING, C., MARSH, S. H. & GILES, J. R. A. (eds) *Elevation Models for Geoscience.*
Geological Society, London, Special Publications, **345**, 5–15.
DOI: 10.1144/SP345.2   0305-8719/10/$15.00 © The Geological Society of London 2010.

addresses the strength of the human perceptual system. Satellite technology has produced vast amounts of remote sensing data that are often understood first, and commonly understood best, by visual interpretation. Over the past four decades, most of these data have been spatially two-dimensional. But the Earth's surface is three-dimensional (3D). Detailed topographical data provide the means to visualize and analyse current, future and archival remote sensing data, within their natural 3D structure, facilitating greater understanding of the features and processes that these data record.

Given all of its uses, demand for elevation data is very high. Consequently, NASA, working with interagency and international partners, has produced (and is continuing to develop) two major contributions to global elevation measurement at 1 arcsecond (or a few arcseconds) spatial resolution (30–100 m). These are the Shuttle Radar Topography Mission (SRTM) and the Advanced Spaceborne Thermal Emission and Reflection Radiometer (ASTER) mission.

SRTM and ASTER

One of the most practical and valuable returns from the United States space programme is the SRTM digital elevation model (DEM). Until the production of the SRTM DEM, good-quality measurements of the Earth's surface at practical levels of detail did not exist or were not generally available for much of the planet. SRTM was developed at NASA's Jet Propulsion Laboratory (JPL) as a joint venture of NASA, the United States National Geospatial-Intelligence Agency (NGA), and the German and Italian Space Agencies (Farr et al. 2007). The mission collected 12 terabytes (10^{12} bytes) of data over nearly all of Earth's landmass between 60°N and 56°S in just 11 days in February 2000. Elevation measurements were derived from interferometric analysis of the C-band radar signal and were processed at JPL. The resultant DEM has 1 arcsecond (c. 30 m) postings, with an absolute vertical resolution significantly better than the mission specification of 16 m (Rodriguez et al. 2005). The SRTM DEM is now freely available (at a somewhat reduced effective resolution for non-US areas). However, the DEM is not spatially comprehensive. It did not cover areas within 30° latitude of the poles and, more troublesome for most users, it has substantial gaps ('voids') where the radar interferometric system failed to provide a signal adequate for DEM generation.

Meanwhile, generally coincident with the SRTM Project, but continuing to 2010 and beyond, ASTER has been acquiring imagery across all areas of the planet up to within 8° latitude of the poles. ASTER is one of the sensors operating on Terra, a satellite launched in December 1999 as part of NASA's Earth Observing System (EOS) (Abrams 2000). The ASTER Project is a co-operative effort between NASA, Japan's Ministry of Economy, Trade and Industry, and Japan's Earth Remote Sensing Data Analysis Center. ASTER covers a wide spectral region with 14 bands from the visible to the thermal infrared, with high spatial, spectral and radiometric resolution. The spatial resolution varies with wavelength: 15 m in the visible and near infrared (VNIR 0.55–0.80 μm); 30 m in the short wave infrared (SWIR 1.65–2.4 μm); and 90 m in the thermal infrared (TIR 8.3–11.32 μm). An additional band is the key to producing digital elevation models. This band (named 3B) is the same as nadir band 3 (NIR), except that it observes at a backward angle of c. 28°, producing a stereo pair for each daytime ASTER image (Welch et al. 1998; Hirano et al. 2003). Each ASTER scene covers an area of 60 × 60 km, and the sensor has up to 8.55° of pointing capabilities. Standard DEMs produced by the United States Geological Survey Eros Data Center (USGS-EDC) have 30 m postings, similar to SRTM's 1 arcsecond postings. However, users can also produce their own DEMs from the band 3 stereo pair using any chosen software. ASTER DEMs are comparable in resolution to those from SRTM. However, potential improvements are still possible since the DEMs do not capture all of the topographic detail that is visually apparent in the stereo imagery.

Topographical exploration

We are in the golden age for the exploration of Earth's surface via satellite data visualization. After a quarter century of high-quality satellite image acquisitions, the production of near-global elevation measurements, and access to these datasets via advanced computers, software and networks, Earth exploration is available to most people with tools as simple as Google Earth™. SRTM provided much of the Google Earth DEM and it complements the resolution of Landsat, the primary satellite imagery. Such merged image–DEM perspectives (Fig. 1) and fly-through visualizations work well even when the imagery is somewhat more detailed than the DEM because the image detail often extends topographical visual cues to higher spatial frequencies, primarily via topographical shading.

Sometimes, however, exploring Earth's surface with pure geomorphic (DEM only) data and user-selected enhancements is especially effective (Figs 2 & 3). Satellite images problematically convolve and obscure topographical shading with land

Fig. 1. Mount Ararat and Little Ararat in easternmost Turkey. Landsat image on SRTM elevation model, near-horizontal southerly view, 1.25× vertical exaggeration. Seismic, volcanic and mass-wasting hazards are all evident in these datasets, and all contributed to a major natural disaster here in 1840 (PIA03399 at http://photojournal.jpl.nasa.gov).

cover reflectance such that these two largely independent variables are not readily distinct. DEMs, of course, measure only the shape and not the radiance of the surface, and so avoid this problem. SRTM provides the best single source of near-global elevation data for pure geomorphic observation.

Stereoscopic satellite views also avoid the problem, but do so instead by perceptual deconvolution (rather than quantitative extraction) of the

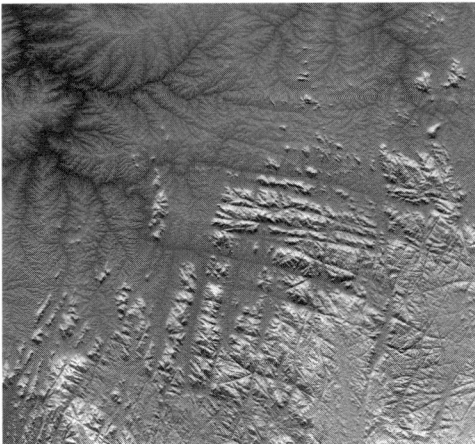

Fig. 2. Lithology and landscape evolution, Gotel Mountains, Nigeria and Cameroon. SRTM DEM mix of shading and height as brightness. Rectangular and other linear drainage patterns in the highlands contrast greatly with the dendritic drainage patterns in the lowlands. These differing geomorphic patterns strongly indicate substantial differences in rock type (PIA04954 at http://photojournal.jpl.nasa.gov).

topographical information from reflectance information. ASTER provides one of the most readily available near-global sources of high-resolution stereoscopic imagery. Significantly, these stereo images reveal topographical detail much finer and more accurate than the DEMs derived from them. This is because individual pixels can be perceived stereoscopically, but each DEM measurement is generated from an areal correlation and is thus somewhat spatially averaged. The standard ASTER DEM uses a 9×9 pixel (135×135 m) kernel, which degrades the DEM spatial resolution to some value much greater than the 15 m pixel size and 30 m posting but somewhat less than the kernel size (c. 120 m).

Synthetic stereo is a simple yet effective method for viewing elevation models, whether incorporating image overlays or just using shading of the DEM itself. Imagery, of course, must first be spatially registered to the DEM. Alternatively, DEM shade images have inherently perfect registration. The synthetic stereo algorithm simply shifts image pixels left for the right-eye image and right for the left-eye image as a linear function of elevation. Shade and other grey image results can be displayed as a red (left eye) and blue–green (right eye) anaglyph, with the use of red–cyan anaglyph glasses, and can be interactively enlarged and roamed on a computer display. Static displays, including full colour displays, can be viewed instead with stereoscopes or without glasses using wall-eyed (parallel) or cross-eyed viewing. Cross-eyed viewing is generally easier than wall-eyed viewing because eyes naturally focus close when they cross. Figure 4 provides an example of a DEM viewed in its full three dimensions, without special glasses, when observed with cross-eyed stereo.

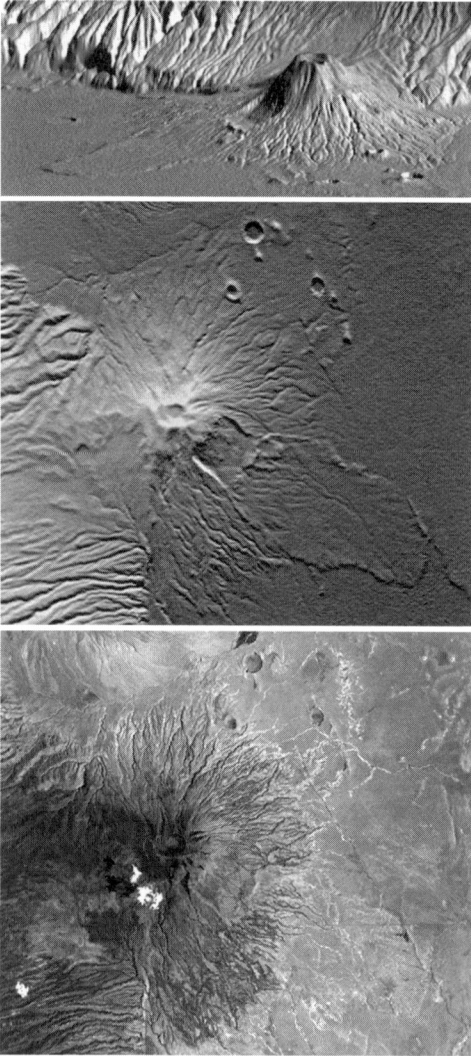

Fig. 3. Crater Highlands, East African Rift, Tanzania. Top: Perspective view of shaded SRTM DEM (PIA06669 at http://photojournal.jpl.nasa.gov). Middle: Nadir view of shaded SRTM DEM, north at top. Bottom: Corresponding Landsat nadir view. Note that the collapse of the SE flank of the volcano and the 10 km-long (and up to 45 m-thick) debris field are clear in the DEM but not recognizable in the Landsat image.

Topographical exploration of Earth has numerous specific uses. A particularly interesting use is the search for interplanetary analogues, especially for Mars (Fig. 5). Mars has no apparent fluid or biotic land cover. All surfaces are petrological (including ice), and globally deposited dust creates a relatively uniform spectral reflectance (except for ice). This near uniformity of reflectance on Mars (at least compared to Earth) makes Mars satellite imagery appear more like Earth shaded elevation models than like Earth satellite imagery. Consequently, in some aspects, Earth exploration for Mars geomorphic analogues may be more readily achieved with SRTM (and ASTER and other) elevation models than with satellite images.

DEM fusion: improving the global DEM

Most users of elevation information view it as a temporally static spatial variable, but certainly an important one that greatly impacts surface and near-surface natural processes. As such, many researchers require elevation data, but without regard to its date of measurement. Typically, they would prefer to simply acquire the best available topographical data rather than generate it or refine it themselves, site by site. Consequently, global fusion of the SRTM and ASTER DEMs into an enhanced and readily accessible standard product is a goal of ongoing work at JPL.

Fundamental differences in the methods of acquisition for SRTM (radar interferometry) and ASTER (photogrammetry) mean that the limitations of each are not highly correlated spatially. In other words, the strengths of each combine synergistically. Clouds are a problem for ASTER but were not for SRTM. Terrain that is either very steep or very smooth has posed challenges for each sensor but in different ways and, therefore, in somewhat different locations.

SRTM elevation data are of reliably high quality but very commonly have voids (areas of missing data). Generally, voids are most common in very steep terrain where the side-looking radar-imaging geometry was problematic, and also in very smooth areas where little of the radar signal was reflected back towards the sensor. Consequently, the locations most impacted by data gaps in the SRTM elevation model are rugged mountains and desert plains and sand sheets. Void filling by interpolation is generally unsatisfactory except for the smallest voids, and the voids can be a hindrance to nearly every use of these data. Filling the SRTM DEM voids with ASTER elevation measurements is an obvious possible solution.

A very simple, yet very effective, method of filling an SRTM DEM void with DEM data from an alternative source was developed by Grohman *et al.* (2006) and was applied in Figure 6 using an ASTER DEM. In simple terms, the method calculates the difference between the surfaces (simple subtraction, but retaining voids), interpolates this 'difference image' across the SRTM void and then adds the result to the alternative (e.g. ASTER)

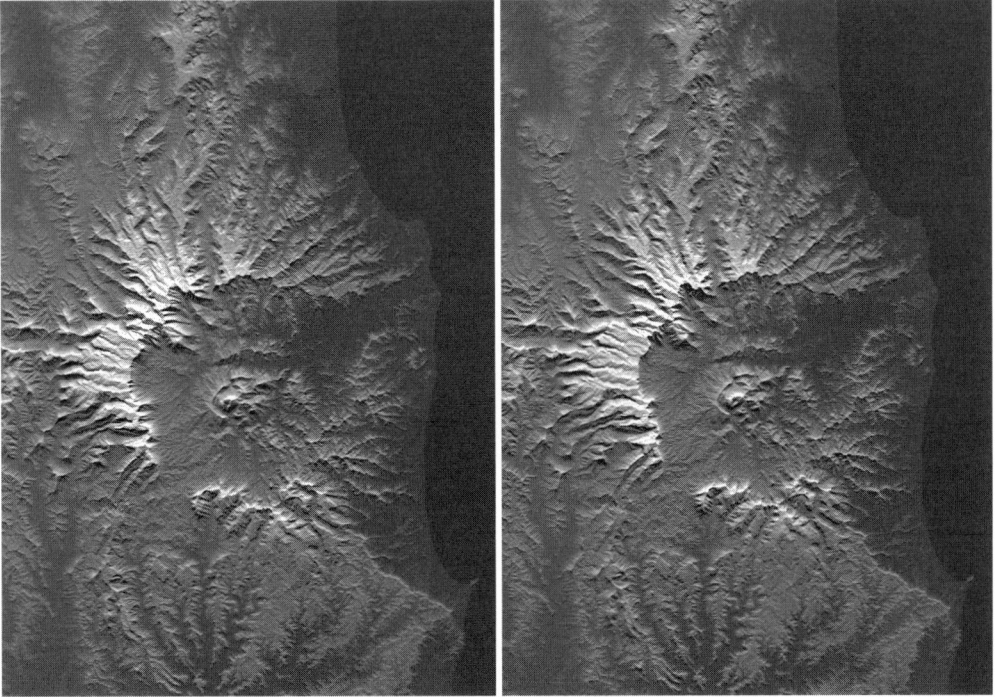

Fig. 4. Tweed Volcano (extinct), Gold Coast, Australia, cross-eyed stereo pair, SRTM shading combined with height as brightness. Area shown is 74 × 102 km (PIA06664 at http://photojournal.jpl.nasa.gov).

DEM. In the resultant merged DEM, SRTM non-void values remain unchanged and the DEM patch is smoothly rubber-sheeted across the void while retaining its relative shape.

ASTER has acquired more than one million scenes. Approximately 45 000 scenes are required to cover Earth's land surface with minimal overlap, but repeat coverage is needed for temporal studies and cloud avoidance. Global daytime coverage is nearly complete and repetitive for most areas, although persistent clouds remain problematic at some locations. (Night-time thermal global coverage is expected too, but early acquisitions were concentrated in southern Asia, and other high relief locations, and along the Antarctic coast.) Since 2006 new software has produced a much improved standard ASTER DEM product but some difficult areas still result in gross errors. Errors occur most commonly on north-facing slopes, due to the viewing geometry of the stereo pair, and over radiometrically smooth terrains and land covers (and large shadows) where photogrammetric pattern matching is difficult.

Recently, an ASTER Global DEM (GDEM) has been produced from the entire ASTER image archive. This project was designed, proposed, and implemented by Sensor Information Laboratory Corporation (SILC), a Japanese company that also produced the software for the new ASTER DEM standard product. GDEM benefits from both cloud masking and multi-DEM averaging, and greatly eases the comparison and merger of ASTER elevation information with that of SRTM. An enhanced version of GDEM is now in production, using additional (recently acquired) scenes, better error corrections, and a smaller (5 × 5) correlation kernel for potentially finer resolution.

The SRTM DEM, even the 3 arcsecond version, is generally of higher quality than individual-scene ASTER DEMs (Fig. 7), and preliminary evaluations of ASTER Global DEM test sites show that the (non-void) SRTM DEM is still generally superior, but not greatly so. The general plan is, therefore, to use SRTM DEM values wherever available, and to use ASTER DEM values to fill voids and other areas not covered by SRTM.

Problem areas will remain even after fusion of the SRTM and ASTER global DEMs, and development of a definitive global elevation model will be an ongoing process using additional and forthcoming data sources and innovative techniques. The global ASTER image archive may contribute

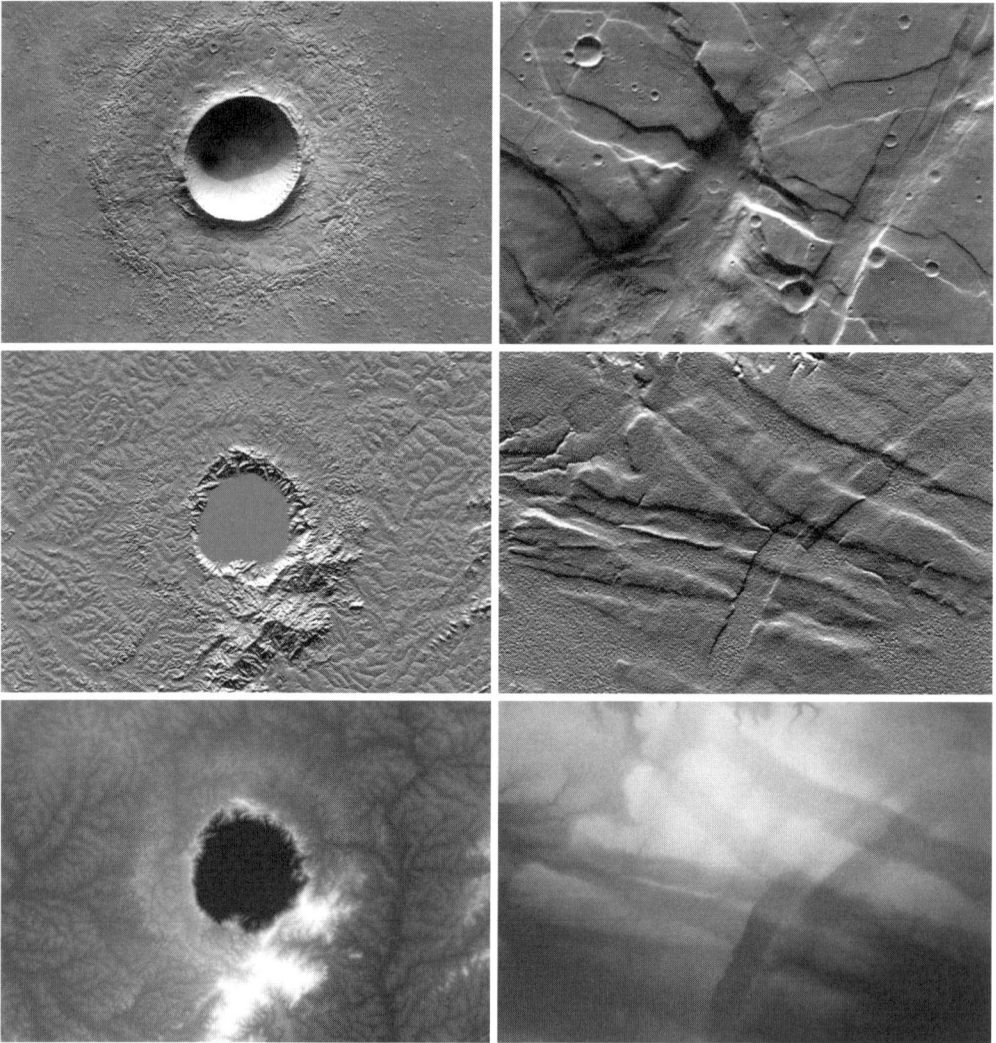

Fig. 5. Analogues of Mars landforms using the SRTM elevation model. Top left: Mars Global Surveyor image of impact crater on Elysium Planitia (PIA02084 at http://photojournal.jpl.nasa.gov). Middle left: Shaded SRTM view of Bosumtwi Crater, Ghana. Bottom left: Bosumtwi Crater, SRTM height as brightness; note especially the ejecta blanket, which is *c*. 35 m thick. Top right: Mars Odyssey image of crossing grabens on Tempe Terra (PIA04471 at http:// photojournal.jpl.nasa.gov). Middle right: Shaded SRTM view of crossing grabens in Afar Triangle, Ethiopia. Bottom right: Afar Triangle grabens, SRTM height as brightness.

to that effort beyond NIR (band 3) photogrammetry. Crippen *et al.* (2007) demonstrated the derivation of elevation values from night-time thermal ASTER images for high-relief terrain in certain environments via the environmental lapse rate. Kirk *et al.* (2005) developed a method of extracting quantitative topographical information from combinations of visible and thermal imagery that may be applicable to ASTER data in some locations. Carlotto

(2000) described a method of enhancing the spatial resolution of a relatively low-resolution elevation model using a relatively high-resolution multispectral image via multidimensional empirical relationships between spectral responses and terrain slopes and azimuths. In addition, Levin *et al.* (2004) determined the topography of sand dunes using shade information from two Landsat non-stereoscopic images with differing sun zenith and

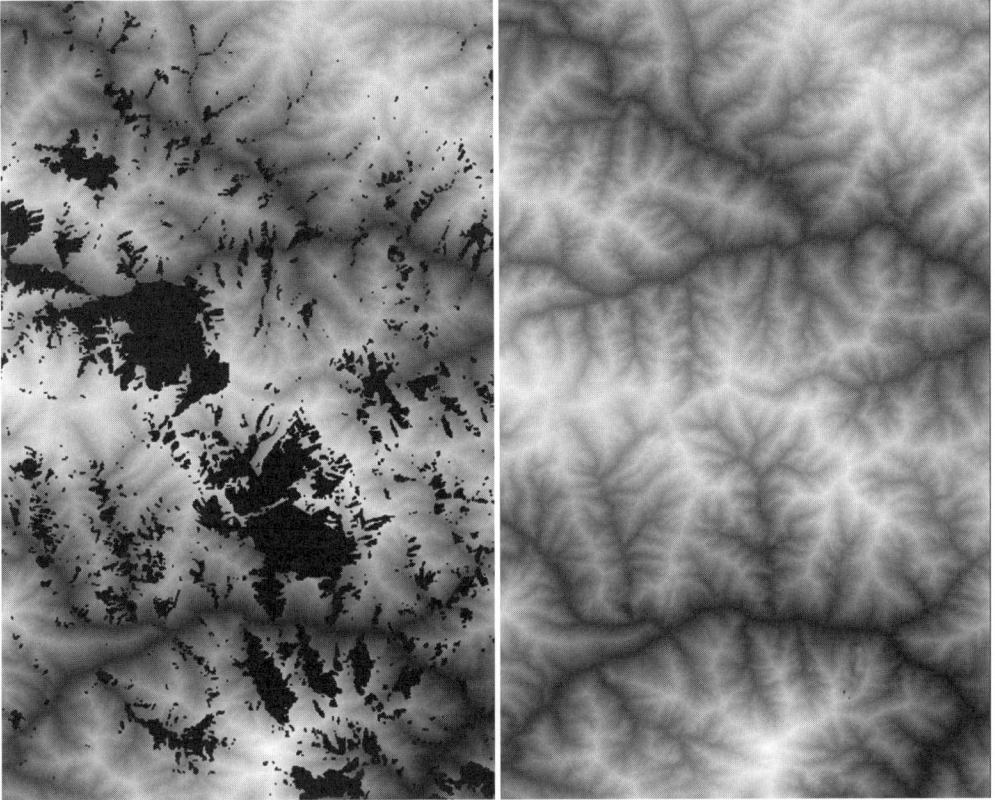

Fig. 6. SRTM voids filled with ASTER DEM, Sichuan Province, China. Height as brightness. Area shown is 41 × 78 km.

azimuth angles (and image resolutions similar to ASTER). This method is called 'photometric stereo'. Notably, they concluded that their result was better than a DEM produced from an ASTER stereo pair.

Indeed, ASTER imagery contains topographical information at resolutions up to the 15 m resolution of the VNIR bands (e.g. band 3, Fig. 7). This level of detail can be seen radiometrically (as natural shading) and stereoscopically but is not now extracted photogrammetrically. Innovative extraction methods might tap this unrealized potential.

DEM differentiation: measuring topographical change

Although topography is essentially static at most temporal and spatial scales of interest at most locations, and for most users' purposes, dynamic topography and its hazards are important in geological studies and land-use planning. Earthquakes, volcanoes, landslides, and extreme erosion and deposition events all produce significant, problematic and even dangerous topographical change. Likewise, glaciers, as part of the solid Earth, exhibit topographical changes that may collectively indicate ominous global climate change.

Elevation differencing is fundamentally a simple subtraction, but spatial registration is critical (Van Niel *et al.* 2008), and systematic differences of the DEMs must not be confused with temporal differences of the surface they were meant to measure. It is generally intended that DEMs measure the interface of rock, soil, ice, lakes and rivers (below the interface) with the atmosphere and above-ground vegetation and buildings (above the interface). Although classic methods of field surveying and aerial photogrammetry have generally excluded above-ground vegetation and buildings while manually mapping the surface, automated satellite methods generally cannot do so. Instead, both SRTM and ASTER map a 'reflectance' surface that includes the vegetation and buildings. Consequently, temporal elevation changes will include vertical land cover changes. These may be interesting signals for ecologists and other researchers

Fig. 7. Resolution and quality comparison of SRTM elevation models and the ASTER image and elevation model, Sichuan Province, China, north at top, 12 × 21 km area. SRTM 1 arcsecond (1 AS), 3 arcsecond (3 AS) and ASTER DEM shown with simulated illumination from the SE. ASTER band 3 (B3) has 15 m resolution and natural illumination from the SE.

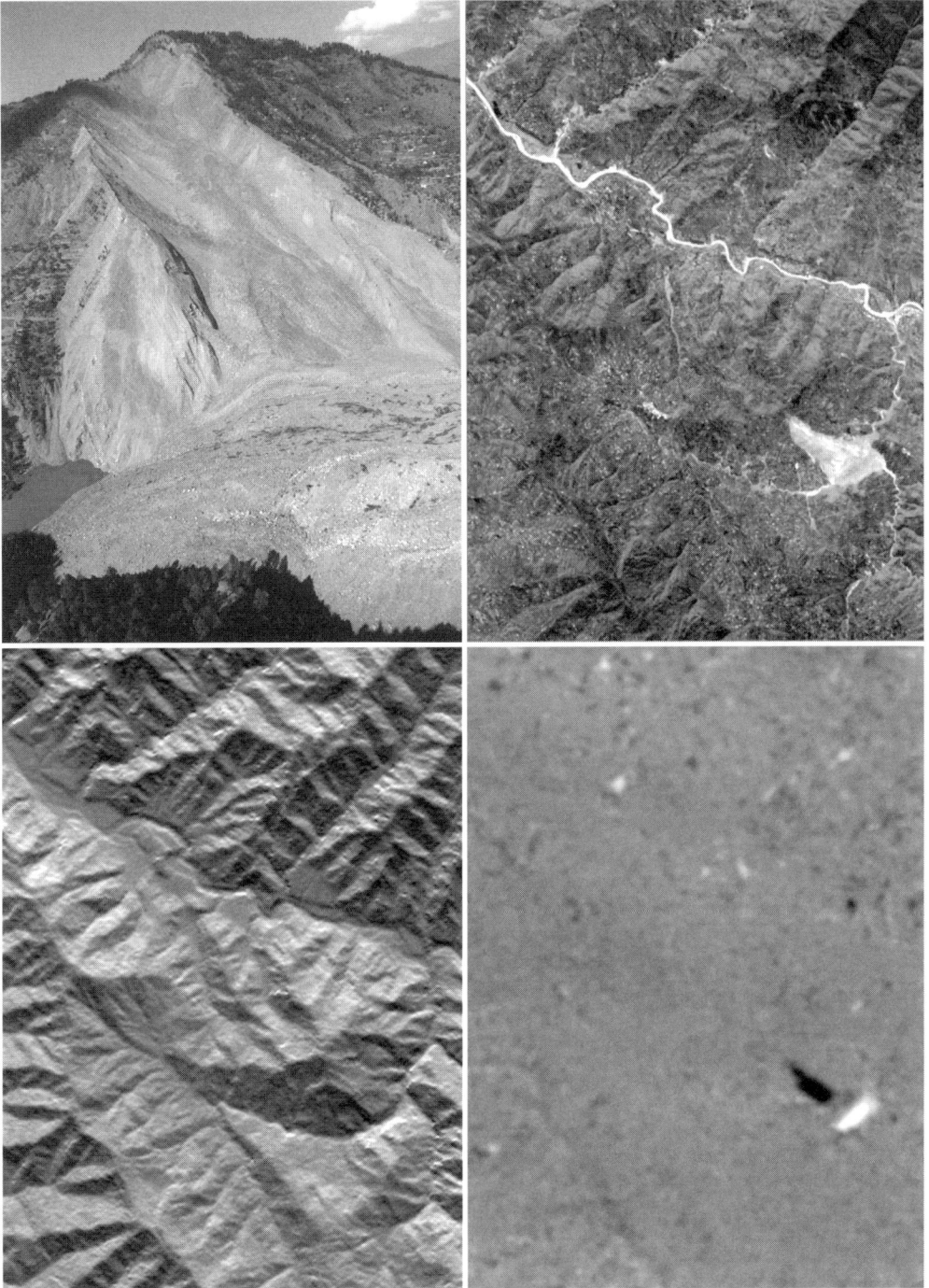

Fig. 8. Hattian landslide in Kashmir triggered by the major earthquake of 8 October 2005. Top left: Photograph (from helicopter) looking NW. Top right: ASTER image difference of bands 1 and 3 (green band minus near-infrared band) showing the landslide scar as bright, indicating a lack of vegetation. The area shown is 11 × 16 km. Bottom left: Corresponding shaded SRTM DEM (pre-quake). Bottom right: Corresponding difference of ASTER pre- and post-quake DEMs shown as bright (up) and dark (down).

(Kellndorfer *et al.* 2004) but they are noise for geologists. Furthermore, radar (SRTM) and near-infrared (stereoscopic ASTER) radiation may reflect from somewhat different levels of a vegetation canopy resulting in a 'systematic noise' in differencing the surfaces they detect. Such issues are important where the signal to noise ratio of topographical change is relatively small.

Elevation change detection for measurement of glacial thinning adds the critical third dimension to satellite surveys when estimating changes in glacial mass that may relate to climate change and sea-level rise (Rignot *et al.* 2003; Rivera *et al.* 2005). The value of such fine measurements critically depends on their accuracy, about which there is currently considerable debate and controversy. Berthier *et al.* (2006) claimed a well-documented bias in SRTM measurements for their study site at Mont Blanc in the French Alps, with elevation underestimated by as much as 10 m at high altitudes. Kääb (2005) found SRTM data to be 7 m too high for a glacial site in the Swiss Alps. Meanwhile, Carabajal & Harding (2006) found variable biases and standard deviations for sites in the western USA and Central Asia when comparing SRTM data with measurements from ICESat LiDAR (Light Detection And Ranging) profiles. Clearly, a better general understanding of SRTM (and ASTER) accuracies and precisions is needed in order to calibrate important findings of small but measurable topographical changes.

Larger topographical changes are less sensitive to the foregoing issues as the change signal is large while the noise remains small. For example, ASTER DEMs and the SRTM DEM of Kashmir were used for volumetric measurements of a major landslide, named the Hattian landslide, and the 248 m-tall natural dam that it created in the major earthquake of 8 October 2005 (Fig. 8). The hazard potential of this site regarding lake growth, possible failure of the landslide dam, and possible generation of an extraordinarily large and catastrophic debris flow was monitored with a series of ASTER images and DEMs. One test used two ASTER DEMs that differed by about 5 years in total but differed in season by only 18 days. The landslide scar (elevation down) and landslide dam (elevation up) are clear relative to the nearby DEM-difference noise of only about 8 m vertically, as viewed in a DEM difference image (Fig. 8, lower right). The 'down' and 'up' volumes are similar, at about 75×10^6 m^3. A difference measurement using the SRTM DEM as the pre-quake DEM provided similar results. Note, however, that the actual landslide volume exceeds the difference measurements because some slide debris remains in the source area.

It is noteworthy that the 'static' topography in areas surrounding the Hattian landslide provides evidence of previous landslides, primarily as hillside scars and dissected terraces of valley-fill deposits that must have accumulated behind other natural dams that are now eroded away. This site provides an excellent example of elevation data exploration revealing past natural processes while also quantifying similar current natural processes.

Conclusion

At 30×30 m resolution, DEM coverage of Earth's landmass involves about 165×10^9 spatially distinct elevation measurements. NASA's SRTM and ASTER missions have contributed to measuring a large majority of the landmass at resolutions approaching 30 m, but much work remains. Merging these two elevation datasets will be highly beneficial for many users. In addition, some void filling, resolution improvement and error correction may be possible using additional information from the ASTER multispectral imagery. These latter efforts might take great advantage of empirical relationships between the images and the DEMs within local areas.

For several years now, the SRTM and ASTER DEMs have provided new views and measurements of our environment that bear upon our understandings across numerous scientific disciplines. In many areas they have provided the first good look at the true 3D nature of Earth's surface. Meanwhile, multitemporal ASTER DEMs, ASTER DEMs with SRTM data, and either of these datasets with historic topographical data have provided some direct measures of geomorphic change. Importantly, they also provided a near-global, near-synoptic baseline for measuring future topographical change.

This research was carried out at the Jet Propulsion Laboratory, California Institute of Technology, under a contract with the National Aeronautics and Space Administration. Mention of commercial products and vendors does not imply endorsement. The Hattian landslide photograph was taken by W. Thompson and acquired via R. Yeats.

References

ABRAMS, M. 2000. The Advanced Spaceborne Thermal Emission and Reflection Radiometer (ASTER): data products for the high spatial resolution imager on NASA's Terra platform. *International Journal of Remote Sensing*, **21**, 847–859.

BERTHIER, E., ARNAUD, Y., VINCENT, C. & REMY, F. 2006. Biases of SRTM in high-mountain areas: implications for the monitoring of glacier volume changes. *Geophysical Research Letters*, **33**, L08502, doi: 10.1029/2006GL025862.

CARABAJAL, C. & HARDING, D. 2006. SRTM C-band and ICESat laser altimetry elevation comparisons as a function of tree cover and relief. *Photogrammetric Engineering and Remote Sensing*, **72**, 287–298.

CARLOTTO, M. 2000. Spatial enhancement of elevation data using a single multispectral image. *Optical Engineering*, **39**, 430–437.

CRIPPEN, R., HOOK, S. & FIELDING, E. 2007. Nighttime ASTER thermal imagery as an elevation surrogate for filling SRTM DEM voids. *Geophysical Research Letters*, **34**, L01302, doi: 10.1029/2006GL028496.

FARR, T., CARO, E. *ET AL.* 2007. The Shuttle Radar Topography Mission. *Reviews of Geophysics*, **45**, RG2004, doi: 10.1029/2005RG000183.

GROHMAN, G., KROENUNG, G. & STREBECK, J. 2006. Filling SRTM voids: the delta surface fill method. *Photogrammetric Engineering and Remote Sensing*, **72**, 213–216.

HIRANO, A., WELCH, R. & LANG, H. 2003. Mapping from ASTER stereo image data: DEM validation and accuracy assessment. *ISPRS Journal of Photogrammetry and Remote Sensing*, **57**, 356–370.

KÄÄB, A. 2005. Combination of SRTM3 and repeat ASTER data for deriving alpine glacier flow velocities in the Bhutan Himalaya. *Remote Sensing of Environment*, **94**, 463–474.

KELLNDORFER, J., WALKER, W. *ET AL.* 2004. Vegetation height estimation from Shuttle Radar Topography Mission and National Elevation Datasets. *Remote Sensing of Environment*, **93**, 339–358.

KIRK, R., SODERBLOM, L., CUSHING, G. & TITUUS, T. 2005. Joint analysis of visible and infrared images: a 'magic airbrush' for qualitative and quantitative topography. *Photogrammetric Engineering and Remote Sensing*, **71**, 1167–1178.

LEVIN, N., BEN-DOR, E. & KARNIELI, A. 2004. Topographic information of sand dunes as extracted from shading effects using Landsat images. *Remote Sensing of Environment*, **90**, 190–209.

RIGNOT, E., RIVERA, A. & CASASSA, G. 2003. Contribution of the Patagonia Icefields of South America to sea level rise. *Science*, **302**, 434–437.

RIVERA, A., CASASSA, G., BAMBER, J. & KÄÄB, A. 2005. Ice elevation changes of Glaciar Chico, southern Patagonia, using ASTER DEMS, aerial photographs and GPS data. *Journal of Glaciology*, **51**, 105–112.

RODRIGUEZ, E., MORRIS, C. & BELZ, J. 2006. A global assessment of the SRTM performance. *Photogrammetric Engineering and Remote Sensing*, **72**, 249–260.

VAN NIEL, T., MCVICAR, T., LI, L., GALLANT, J. & YANG, Q. 2008. The impact of misregistration on SRTM and DEM image differences. *Remote Sensing of Environment*, **112**, 2430–2442.

WELCH, R., JORDAN, T., LANG, H. & MURAKAMI, H. 1998. ASTER as a source for topographic data in the late 1990s. *IEEE Transactions on Geoscience and Remote Sensing*, **36**, 1282–1289.

Refinement of ASTER digital elevation models using SRTM data for an environmental study in China

M. R. A. NIGHTINGALE[1]* & G. L. K. MORGAN[2]

[1]*Rushmere Folly, 5 Thornley Drive, Ipswich, Suffolk IP4 3LR, UK*

[2]*Department of Earth Science and Engineering, Imperial College London,
Prince Consort Road, London SW7 2AZ, UK*

Corresponding author (e-mail: mark.nightingale@uclmail.net)

Abstract: This paper presents a simple but effective method to improve Advanced Spaceborne Thermal Emission and Reflection Radiometer (ASTER) digital elevation models (DEMs). Digital elevation models are an important component of geomorphological modelling so their integrity is vital to achieve reliable results. For this investigation the relative ASTER DEM produced from an automated cross-correlation algorithm was not considered accurate enough, so 3 arcsecond (90 m) Shuttle Radar Topography Mission (SRTM) data were used to modulate the ASTER DEM.

The method presented in this paper allows the SRTM to enforce vertical control on the relative ASTER DEM whilst attempting to maintain the ASTER DEM's 30 m spatial resolution. The process is fast and efficient, and can be applied to other DEMs. There is, however, a compromise since the fusion process, involving the averaging of the values, does potentially mean that some of the detail in the original 30 m ASTER DEM will be lost.

The Three Gorges Dam on the Yangtze River is the world's largest industrial project and has been designed to solve the problem of flooding, generate a large amount of hydroelectric power and improve navigation for shipping up the Three Gorges. One of the characteristics associated with the Yangtze River is the huge volume of sediment carried from the upper sources (Jinsha River) to the river mouth at Shanghai. This sediment, liberated through the processes of erosion, presents a problem to the lifespan of the dam. The problem of erosion in the Jinsha River area is extremely severe because of a combination of natural and human factors. To implement soil conservation and slope stability measures it is essential to investigate and monitor erosion processes in the primary source of sediments in the Jinsha River catchments.

Considerable effort has been undertaken to monitor erosion in an effort to reduce land degradation and environmental damage. This work was initially undertaken using *in situ* assessments; however, it was often costly, time-consuming and not practical in remote areas. The development of aerial surveys and subsequent satellite remote sensing has drastically improved erosion assessments, offering continuous coverage of large areas with a high spatial resolution in an often cost-effective manner (Morgan 2005). Multispectral scanners provide information on geology, soils and also vegetation state. Of equal importance is

surface elevation data provided by satellite-borne sensors and the derivative measurements of slope and aspect, which are essential parameters for geomorphological modelling (Welch *et al.* 1998). One of the most widely used sources of digital topographical data is from the Shuttle Radar Topography Mission (SRTM). This is a joint project between the National Aeronautics and Space Administration (NASA) and the National Imagery and Mapping Agency (NIMA). The objective was to collect radar data, which were then interferometrically processed, to produce digital topographical data for 80% of the Earth's land surface (from 60°N to 56°S). All the data were acquired using single-pass interferometry (with the two radar antennas separated by a 60 m boom) in a single mission lasting 11 days, scanning the Earth's surface independent of darkness or cloud cover (Rabus *et al.* 2003). It is considered to be the most complete high-resolution topographical dataset of the globe (Obrock & Guth 2005). For areas outside the United States the DEMs are provided in geographic co-ordinates with a horizontal resolution of 3 arcsecond (90 m). The absolute vertical accuracy is ±16 m (linear error at 90% confidence, LE90) (Kääb 2005). WGS84 is the horizontal and vertical datum. Unfortunately, this horizontal resolution is coarse, thus making the international continental SRTM data a good tool for regional-scale environmental research but limiting its use for small-scale

From: FLEMING, C., MARSH, S. H. & GILES, J. R. A. (eds) *Elevation Models for Geoscience.*
Geological Society, London, Special Publications, **345**, 17–21.
DOI: 10.1144/SP345.3   0305-8719/10/$15.00 © The Geological Society of London 2010.

studies. Digital elevation models can also be produced by multipass interferometry from satellite-based SAR (synthetic aperture radar) systems such as ENVISAT or ERS, which can generate high-accuracy ($< \pm 10$ m) Z co-ordinates (Welch *et al.* 1998). The interferometric processing is not trivial and potentially less accessible because it requires specialist software. Caution has to be taken when generating DEMs from multipass interferometry since the time interval (temporal baseline) between SAR acquisitions can result in low coherence in interferometric SAR (InSAR) processing. Changes in atmospheric conditions between SAR acquisitions can also make multipass InSAR processing more complicated. The main benefit of the SRTM dataset is that the Jet Propulsion Laboratory (JPL) in Pasadena, CA, has processed the radar data into elevations models that are freely available to download from the United States Geological Survey's Earth Resources Observation Center and Science (USGS EROS) Data Center.

For this investigation a DEM was generated using the along-track capabilities of the Advanced Spaceborne Thermal Emission and Reflection Radiometer (ASTER). These DEMs can have a higher horizontal spatial resolution than SRTM (30 m in this case) for regions outside the United States. The 14 spectral bands of ASTER are exactly registered to the ASTER-derived DEM so it makes sense to use ASTER as a base DEM.

The ASTER DEM

ASTER stereo data are recorded by band 3 (0.78–0.86 μm) in the visible and near infrared (VNIR) region, using nadir and rear-viewing telescopes. This subsystem provides a base-to-height ratio of 0.6, which is close to ideal for generating DEMs by automated techniques (Hirano *et al.* 2003). The NASA Terra platform, on which ASTER is mounted, is at a nominal altitude of 705 km and the push-broom linear array covers a 60 km ground swath at a 15 m spatial resolution (for the VNIR imagery). The along-track mode is a major advantage (compared to the cross-track, such as SPOT1–4) for data acquisition because the stereo pairs are acquired 60 s apart under uniform environmental and lighting conditions, resulting in stereo-pairs of consistent quality (Hirano *et al.* 2003). Registering two images to the same ground area recorded from the nadir and rear-viewing telescopes generates the DEMs. Any positional differences parallel to the direction of satellite travel in the stereo pair are attributed to displacements caused by relief (Abrams 2000). Automated stereocorrelation has become a standard method of generating DEMs from digital stereo images; however, the

accuracy of the relative DEM depends directly on the quality of the original elevation data and the algorithms used in photogrammetric software (Cuartero *et al.* 2004). The software used to produce the ASTER DEM in this work is AsterDTM 2.0®, an ENVI 4.1® add-on module created by SulSoft Ltda. An automated cross-correlation method was used to produce a 30 m spatial resolution (1 arcsecond) relative DEM and this method does not require the manual selection of ground control points (GCPs), as it uses the satellite ephemeris data only. The relative DEM accuracy varies according to the literature. Sulsoft Ltda claimed the vertical accuracy (root mean square error (RMSE)) was better than 15 m (Sulsoft Ltda 2003), whilst Abrams (2000) stated the accuracy as being 15–30 m and Hirano *et al.* (2003) as 10–30 m.

For the purposes of the study in China, having an accurate DEM is essential and so Shuttle Radar Topography Mission (SRTM) data are used as an external DEM to modulate the relative ASTER DEM. The vertical accuracy of SRTM is considerably better than that of the relative ASTER DEM, so the SRTM DEM is used to enforce vertical control on the relative ASTER DEM whilst maintaining the ASTER DEM's 30 m spatial resolution.

DEM refinement

On comparing the relative ASTER DEM with the SRTM DEM it was found that the ASTER often has a terracing characteristic forming a contouring effect parallel to the slopes, possibly a limitation of the algorithm used by AsterDTM 2.0® (Fig. 1). When the ASTER elevation values were compared to the topographical maps there was a systematic offset of ± 50 m, with the ASTER values being much lower. When compared with the SRTM elevation data the ASTER DEM values were also much lower than the corresponding elevation values on the SRTM data (Table 1, Fig. 2).

To enforce vertical control on the ASTER DEM a simple fusion formula was initially used that allowed the maximum and minimum values to be dictated by the SRTM data whilst the rest of the ASTER DEM was 'smoothed' by an average of the ASTER and SRTM values. The formula is as follows:

$$\text{if } i_1 > i_2 \text{ then } i_2 \text{ else if } \text{abs}(i_1 - i_2)$$
$$> 600 \text{ then } i_2 \text{ else } (i_1 + i_2)/2$$

where i_1 is the ASTER DEM and i_2 the SRTM DEM.

The value 600 was determined empirically by viewing the DEM with a sun-shading. Patches

(a)

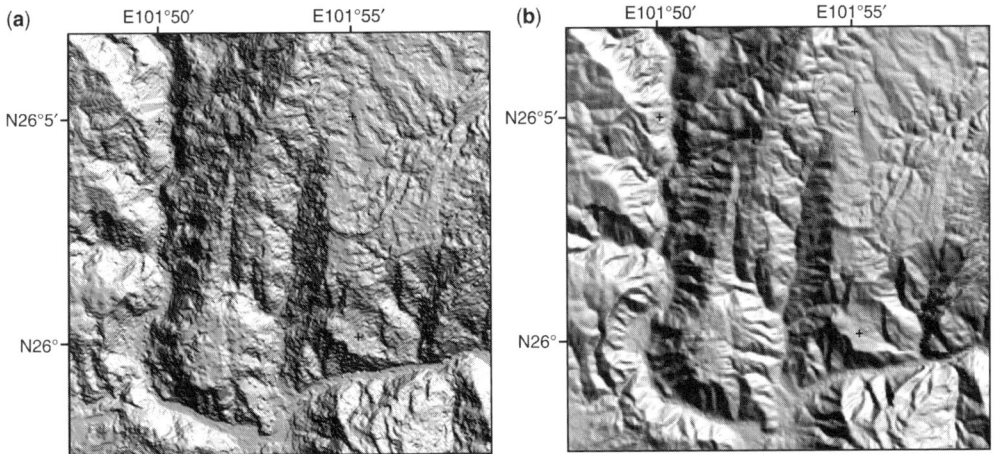

(b)

Fig. 1. (**a**) A subset from the raw ASTER relative DEM produced with the AsterDTM 2.0® module. Note the terracing effect (WGS84 UTM zone 47N). (**b**) The 3 arcsecond SRTM DEM of the same area is a much higher quality DEM, having been processed by NASA JPL, and is smoother and more uniform (WGS84 UTM zone 47N).

appeared when the threshold was not set high enough and the ASTER data had been replaced directly by the SRTM data (Fig. 3). This 'smoothed' ASTER DEM is certainly better, with a reduction in the systematic shift between the relative ASTER DEM and the SRTM DEM (Fig. 2).

A better approach is to use ground control points (GCPs) in the production of the ASTER DEM. The external GCPs calibrate the elevation values, converting the relative ASTER DEM into an absolute one. The SRTM DEM was co-registered to the ASTER scene to the nearest pixel, making it possible to extract the SRTM elevation data at exact pixel locations in the ASTER scene. These SRTM elevation values were then manually input as GCPs into the AsterDTM 2.0® module to give the

Table 1. *Example elevation values from the ASTER and SRTM transects*

Easting (m)	Elevation values (m)		
	ASTER DEM	SRTM DEM	Difference (ASTER−SRTM)
801364	1711	1774	−63
801195	1733	1794	−60
801617	1768	1835	−67
801863	1830	1909	−78
801772	1760	1818	−57
801041	1693	1763	−71
801821	1793	1849	−56
800450	1892	1966	−74

Note: The ASTER values are consistently lower than the SRTM values for the corresponding pixel.

SRTM values at the exact pixel locations in the ASTER image. The GCPs are used within the module to move the output DEM to the correct geolocation in *XYZ* space by means of a rubber-sheet warping (SulSoft Ltda 2003). The accuracy of the DEM will depend on the number of GCPs used; generally, the more GCPs the better. In this case 30 well placed GCPs were chosen, and their exact pixel location was known. The accuracy of the ASTER scene, SRTM co-registration and also the accuracy of the SRTM data will therefore determine the accuracy of the GCPs.

The DEM produced with the GCPs is considerably better than that from the fusion formula used above and virtually removes the offset between the relative ASTER DEM and the SRTM DEM. Transects of the DEM are seen in Figure 2, and it is possible to see the closer correlation to the SRTM DEM of the ASTER DEM produced with the external GCPs compared with that of the original relative ASTER DEM. However, there is still some discrepancy between this DEM and the original SRTM DEM. This is because only 30 GCPs are used to calibrate the whole ASTER DEM, and interpolation is undertaken between the GCPs to correct the geolocation in *XYZ* space. To improve this further the DEM produced with external GCPs was then fused with the SRTM data in a similar process to that above. This allows the ASTER DEM with GCPs to be improved on a pixel-by-pixel basis, thus producing a DEM that correlates very well to the original SRTM DEM (Fig. 2). This is the final DEM produced for this work and the topographical transects in Figure 2 show the close vertical correlation with the SRTM DEM, whilst

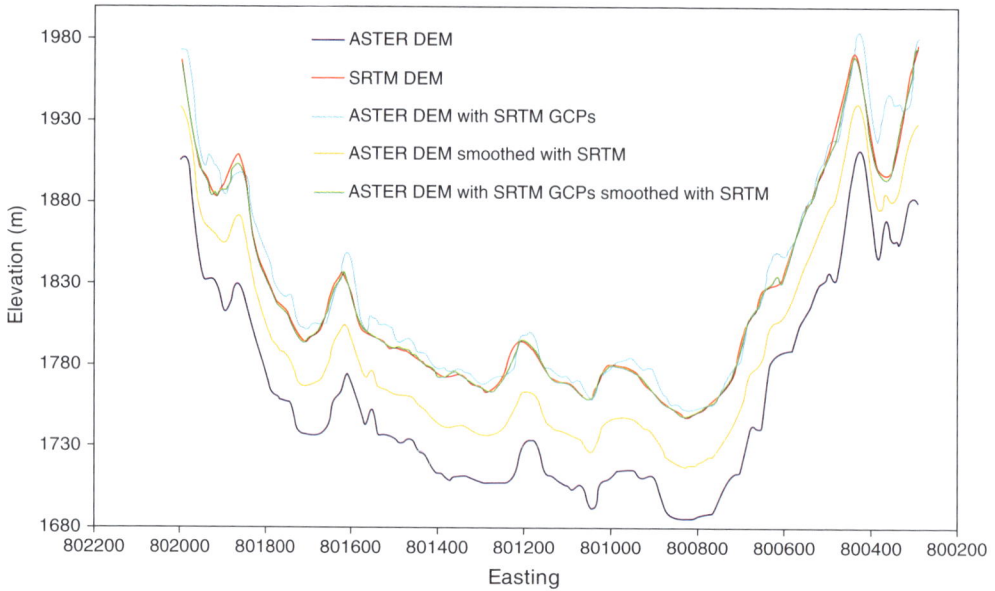

Fig. 2. Traverse through the different DEMs for the study area.

maintaining the high spatial resolution of the original ASTER DEM where possible.

Conclusion

The method outlined above is a simple method to improve the accuracy of an elevation model using another DEM of known accuracy. It also allows a relative DEM to be transformed into an absolute one when it is not possible to attain GCPs from GPS measurements.

The use of GCPs reduces the systematic offset between the SRTM and ASTER DEMs whilst the fusion technique smoothes the ASTER DEM, thus the final fused DEM is composed of average values and some pure SRTM values. These pure SRTM values are assigned to elevation spikes and spurious values by setting a threshold value, above which SRTM values are used. This method does,

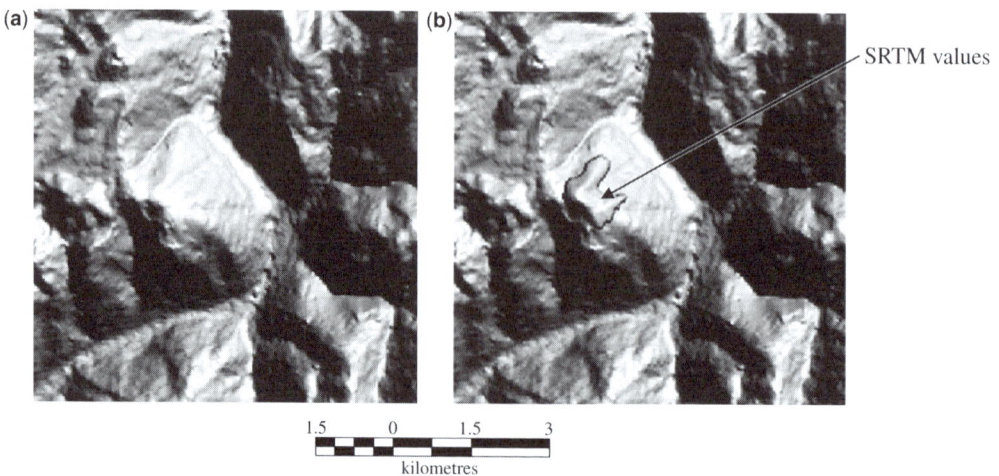

Fig. 3. (a) A subset from the raw fused ASTER and SRTM DEM with the threshold value set at 600. (b) The same subset but with the threshold set at 300. The area marked is where the much higher SRTM data have been used (WGS84 UTM zone 47N).

however, rely on the external DEM (in this case SRTM) being accurate or being of a known and acceptable accuracy.

It must be noted, however, that the process of fusing the ASTER DEM with the SRTM DEM that is replacing values with the mean of the SRTM and ASTER value does average out the greater detail that is in the higher resolution ASTER DEM. It does not eliminate the detail completely but reduces it. This is because 1 SRTM pixel is used to average 9 ASTER pixels. However, this compromise does allow vertical control from an elevation model of known accuracy to be enforced on one of unknown accuracy.

A recent development in the production of ASTER DEMs is new software developed by the Land Processes Distributed Active Archive Center (LP DAAC) and released on 24 May 2006. Since April 2001 they have offered a DEM manually produced from ASTER level-1A data, which was available either as a relative or absolute DEM. The relative DEM was one of their most popular ASTER data products but, unfortunately, LP DAAC became aware of degradation in the accuracy of the ASTER relative DEM over time. An assessment of the accuracy of relative ASTER DEMs produced from a variety of software also prompted the LP DAAC to develop this new software, which allowed for the batch production of relative DEMs with improved accuracy. Validation testing shows the new software produces relative ASTER DEMs more accurate than 25 m RMSExyz (LP DAAC 2006).

References

ABRAMS, M. 2000. The Advanced Spaceborne Thermal Emission and Reflection Radiometer (ASTER): data products for the high spatial resolution imager on NASA's Terra Platform. *International Journal of Remote Sensing*, **21**, 847–859.

CUARTERO, A., FELICISIMO, A. M. & ARIZA, F. J. 2004. Accuracy of DEM generation from TERRA-ASTER stereo data. *In: Proceedings of the XXth ISPRS Congress, 12–23 July, Istanbul, Turkey*. International Society for Photogrammetry and Remote Sensing (ISPRS), Beijing.

HIRANO, A., WELCH, R. & LANG, H. 2003. Mapping from ASTER stereo image data: DEM validation and accuracy assessment. *ISPRS Journal of Photogrammetry & Remote Sensing*, **5–6**, 356–370.

KÄÄB, A. 2005. Combination of SRTM3 and repeat ASTER data for deriving alpine glacier flow velocities in the Bhutan Himalaya. *Remote Sensing of Environment*, **4**, 463–474.

LP DAAC 2006. *Change in the ASTER DEM Production Software at the LP DAAC*. Land Processes Distributed Active Archive Center. Available online at http://edcimswww.cr.usgs.gov/pub/imswelcome/ (accessed 15 July 2006).

MORGAN, R. P. C. 2005. *Soil Erosion and Conservation*, 3rd edn. Blackwell Science, Oxford.

OBROCK, K. & GUTH, P. 2005. Filling holes in SRTM DEMs using ASTER DEMs. *Paper presented at the Proceedings of the 6th International Conference on Military Geology and Geography*. School of Geography, University of Nottingham, UK, 19–22 June.

RABUS, B., EINEDER, M., ROTH, A. & BAMLER, R. 2003. The shuttle radar topography mission – a new class of digital elevation models acquired by spaceborne radar. *ISPRS Journal of Photogrammetry & Remote Sensing*, **4**, 241–262.

SULSOFT LTDA 2003. *AsterDTM 2.0 Installation and User's Guide* [computer software manual]. Available online at: http://www.envi.com.br/asterdtm/english/index.htm (accessed 2 July 2005).

WELCH, R., JORDAN, T., LANG, H. & MURAKAMI, H. 1998. ASTER as a source for topographic data in the late 1990s. *IEEE Transactions on Geoscience and Remote Sensing*, **36**, 1282–1289.

An assessment of ASTER elevation data over glaciated terrain on Pourquois Pas Island, Antarctic Peninsula

A. CZIFERSZKY, A. H. FLEMING* & A. FOX

British Antarctic Survey, High Cross, Madingley Road, Cambridge CB3 0ET, UK

Corresponding author (e-mail: ahf@bas.ac.uk)

Abstract: This paper assesses the accuracy of digital elevation models (DEMs) generated from ASTER (Advanced Spaceborne Thermal Emission and Reflection Radiometer) satellite imagery in glaciated and mountainous terrain. This is an especially attractive technique for inaccessible glaciated regions of the Antarctic Peninsula. Our aims are to determine the best option for generating elevation data from ASTER data and to determine the errors involved for different terrain types. We use a near-coincident and contemporaneous airborne photogrammetric dataset to provide sufficiently accurate reference data. Three different software options for derivation of the ASTER DEM are assessed in order to determine the best available method of generating elevation data for this type of terrain. DEM accuracy is highly correlated to the steepness of the terrain. For areas of low to moderate relief errors are less than 20 m RMSE (root mean square error), but areas of high relief show much larger errors of up to 200 m RMSE.

Access to digital topographical data is essential for a wide range of scientific research and logistics operations in the Antarctic. The high costs and effort involved in airborne and ground-based methods have led to significant effort to deliver elevation data from satellite-based sensors. Until recently, satellite systems such as SPOT have employed across-track sensors to acquire the necessary stereo imagery (Al-Rousan & Petrie 1998). However, the time delay between across-track acquisitions means that acquiring suitable cloud-free images can be difficult, time-consuming and expensive because of bad weather in mountainous areas. The Shuttle Radar Topography Mission (SRTM) will eventually provide access to digital elevation data for large parts of the world and accurate to $c.$ ± 16 m (Farr *et al.* 2000). However, the Polar Regions are not covered by the SRTM dataset and alternative sources of elevation data must be considered.

Modern satellite systems, including SPOT-5 and Advanced Spaceborne Thermal Emission and Reflection Radiometer (ASTER), provide along-track stereo imaging of the land surface. This reduces the problems associated with changes in ground surface and cloud cover between acquisitions of stereo pair images. When compared with other methods the low cost of ASTER imagery (US$80 for Level-1a data including stereo images in 2007) makes this a very efficient way of generating digital elevation data. Satellite radar interferometry can also be used to generate digital elevation models (DEMs), but the loss of coherence over snow and ice surfaces between successive passes means that this technique is not ideally suited to polar environments. Whilst a satellite-based

method cannot currently approach the accuracy of airborne techniques, the errors are likely to be acceptable for a number of applications and represent a significant improvement on what is currently available (Rees 2005) for large portions of the Antarctic continent.

This paper presents the results of work to determine the accuracy of elevation data from ASTER imagery for terrain types typical of the Antarctic Peninsula. Cheng *et al.* (2003) have previously used ASTER data for DEM generation of the Antarctic icesheet, and we are keen to extend the assessment to include the more varied mountainous terrain of the Antarctic Peninsula.

Given that there are a number of software options for generating DEMs from ASTER data, we also assess three of these options. These include the standard ASTER DEM product generated by the USGS EROS Data Center and outputs from two commercial software packages. As we are interested in assessing methods that are applicable to all parts of the Antarctic continent, in each case we evaluate DEMs generated without the use of ground control data, although ground control points (GCPs) were used as an independent check for gross errors. Ground control data are not available or easily obtainable for the great majority of the Antarctic continent owing to the extensive area and the considerable logistical effort required to collect it.

The study area is Pourquois Pas Island on the western Antarctic Peninsula (Fig. 1), which contains a range of terrain types typical of the Antarctic Peninsula including glaciated and steep mountainous terrain. A set of colour aerial photographs together with GPS measured camera frame centres

From: FLEMING, C., MARSH, S. H. & GILES, J. R. A. (eds) *Elevation Models for Geoscience.*
Geological Society, London, Special Publications, **345**, 23–32.
DOI: 10.1144/SP345.4   0305-8719/10/$15.00 © The Geological Society of London 2010.

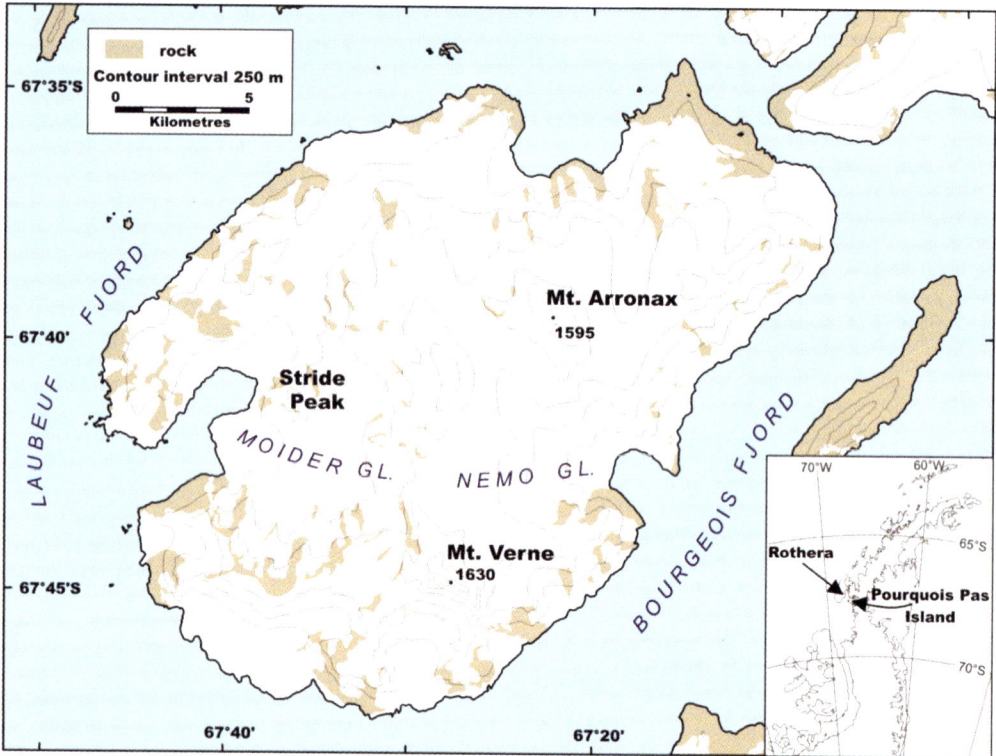

Fig. 1. The location of Pourquois Pas Island, Antarctic Peninsula.

and highly accurate GCPs are available from which we generate control elevation data.

Methods

The overall method we follow for comparing the different DEMs and to draw conclusions about the quality of each satellite-derived DEM is described below.

(1) Identify small (500 × 500 m) and homogeneous sample areas within the study area (Fig. 2) that have typical characteristics (terrain cover, aspect, surface texture, slope) in terms of topography around the Antarctic Peninsula.

(2) Generate satellite-derived DEMs from the ASTER Level-1a data product using three different methods.

(3) Digitize three-dimensional (3D) data points within the sampling areas using the photogrammetric project of the study area as a reference dataset.

(4) Compare the height of the digitized data points to the height value of the corresponding cell in each of the different satellite-derived DEMs by calculating the difference, ΔZ.

Given that the DEM accuracy depends on the cell size, the 30 m pixels used are large. However, this cell size is fixed to achieve valid comparison with the standard NASA ASTER DEM product. We note that as each checkpoint is included in the RMSE calculation, uneven distribution of checkpoints could bias the result. However, emphasis was placed on ensuring an even distribution of the manually digitized checkpoints. Future work will further investigate the effect on the DEM accuracy of altering the DEM output cell size and the options for comparison of DEM pixel values with the digitized photogrammetric point data.

(5) Calculate statistics for all ΔZ within each sample area and for overall sample areas with similar topographical characteristics.

Satellite data

The input ASTER satellite image was acquired on 30 December 2004. The Level-1a product provided by the USGS EROS Data Center is used as the input for all ASTER DEM generation. The ASTER data product includes a nadir and backward-looking

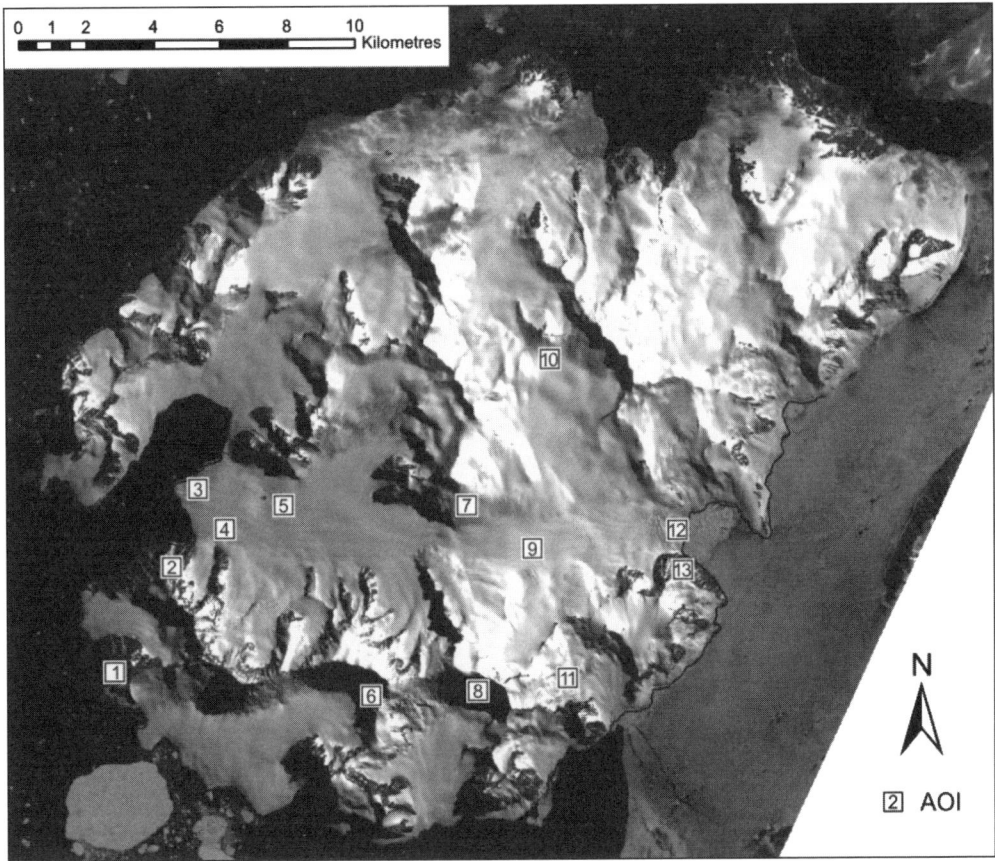

Fig. 2. The Pourquois Pas study area with the 13 AOIs indicated.

view for Band 3 (near infrared) data of the same image area. Positioning of the satellite data and derived DEM data is based purely on the satellite parameters and does not involve the ground control data collected in conjunction with the air photography. Further details of the ASTER satellite data product are available from the USGS EROS Data Centre (EDC 2001).

Options for ASTER DEM generation

The three options for generation of DEM from ASTER satellite data are listed below. In some cases little user input was required, for example, with the NASA product. In other cases, where user input was required, details of the parameters used are included.

(1)　NASA DEM
　　ASTER 14 relative DEM data product, second version, released May 2006 (Lang & Welch 1999).
　　http://lpdaac.usgs.gov/aster/ast14dem.asp

(2)　SILCAST DEM
　　Version 1.07. Commercial software product supplied by Sensor Information Laboratory Corp, Japan.
　　http://www.silc.co.jp/en/products.html
　　User parameters set are detailed below:
  - automatic correction of geoid heights for output DEMs referring to mean sea level (MSL)
  - automatic identification of water bodies

(3)　ITT ENVI DEM module
　　Version 4.3. Commercial software product supplied by ITT Visual Information Solutions.
　　http://www.ittvis.com/envi/addons_DEM-module.asp
　　User parameters set are detailed below:
  - elevation: -5 to $+1855$ m
  - no use of GCPs
  - 60 tie points
  - output pixel size: 30 m
  - minimum correlation: 0.6
  - moving window size: 15×15

- terrain relief: high
- terrain detail: Level 5 (max.)

The Leica Photogrammetry Suite (LPS) (Version 8.7) software was also used to generate a DEM from the ASTER satellite dataset. However, a useable result was not achieved because LPS does not include an ASTER camera model and is primarily a photogrammetric tool not optimized for the generation of elevation data from ASTER satellite data. Options for DEM generation in LPS also require ground control points as input, which is not consistent with our aim to have a method independent of ground measurements.

Reference elevation data

The reference dataset comprises a photogrammetric project consisting of 53 colour images taken with a Zeiss RMK-A aerial survey film camera at a flying height of 4500 m above mean sea level (MSL). The study area is covered by a block of three strips plus some cross-strips that were taken on 20 January 2005. Each image was scanned at a resolution of 20 μm, giving an approximate pixel size of the digital image of 0.6 m at MSL.

Acquisition dates of the ASTER satellite image and reference photography are separated by only 3 weeks. This is of great importance in order to minimize the impact of fresh snow accumulation or snow–ice melt. A separation of several months or years might include significant changes in the surface elevation and characteristics, meaning any comparison would be less relevant.

For the purpose of ground control, the camera frame centre positions at the time of exposure were recorded using kinematic GPS. The GPS sample interval was 1 Hz and the actual camera positions were interpolated within the 1 s sample intervals. The GPS base station was located at Rothera Research Station, *c.* 50 km west of the study area. The processing results for the kinematic GPS, over a 50 km baseline from the GPS base station, showed an RMS accuracy of 0.02 m, and before and after track crossings, on the runway apron, were within 0.05 m in elevation.

In addition to the kinematic GPS data, five GPS GCPs were measured using Trimble GPS receivers of geodetic quality. The measurements were post-processed, resulting in a positional accuracy of better than 0.02 m RMSE in x, y and z, which is much better than the measurement precision from the photography. Although GCPs as well as GPS measured camera frame centres are not strictly required in order to perform a bundle block adjustment, the GCPs are used as independent check points to detect gross errors.

The maximum residuals in the block-adjustment exterior orientations for the 53 photographs were $x = 0.19$ m, $y = 0.13$ m and $z = 0.19$ m. The mean residual at the tie-points was $x = 0.18$ m, $y = 0.20$ m and $z = 0.31$ m, with standard deviations of less than 0.1 m. Thus, the primary block is considered accurate to better than 0.5 m in x, y and z. Therefore, we consider this a suitable reference dataset for assessment of the DEMs derived from the ASTER image.

Sampling areas

A total of 13 sampling areas at a size of *c.* 500 × 500 m were chosen (Fig. 2). Each sampling area, or 'area of interest' (AOI), was chosen to represent an area of homogenous topography. Distribution of the sample areas was also chosen to represent a number of different terrain types typical of the island and the Antarctic Peninsula. This allows us to assess how each ASTER DEM performs for a certain terrain type. The AOIs are chosen to represent differences in terrain steepness, surface texture, aspect and illumination.

Leica's Stereo Analyst software was used to digitize 3D data points from the reference photogrammetric dataset for each AOI. Different numbers of data points were collected for each AOI depending on the height relief, the surface texture and the terrain type. Table 1 gives an overview showing the different AOIs and their characteristics. The AOIs are classified into three main terrain types typical of the glaciated mountainous areas of the Antarctic Peninsula. We expect terrain relief to be the key parameter affecting the results and so have based this classification primarily on surface slope as detailed in Table 1:

(1) low relief glacier surface
(2) moderate relief glacier/icefall
(3) high relief steep rock/ice face

Results and discussion

The vertical accuracy for all digitized points in the AOIs ranges from 75 m RMSE for ENVI to just greater than 90 m RMSE for NASA and SILCAST. In isolation, these figures suggest the method does not meet the design specifications for ASTER DEM data in this region. However, these figures are misleading given that the errors vary so significantly for areas of different slope. It is necessary to separately consider the errors for the three different terrain types outlined previously.

Table 2 shows the mean values of ΔZ and RMSE for the digitized sampling points within the same

Table 1. *Characteristics of the 13 sampling areas*

AOI no.	Terrain type	Terrain description	Aspect	Surface texture	No. of data points	Height difference across the AOI
3	1	Glacier tongue	West	High	4861	Low (49 m)
4	1	Glacier	West	High	1491	Low (37 m)
5	1	Glacier	West	Medium	3810	Low (92 m)
9	1	Glacier	East	Low	70	Low (31 m)
7	2	Icefall	South	High	281	Moderate (347 m)
10	2	Glacier	South	Low	368	Moderate (137 m)
11	2	Icefall	East	High	325	Moderate (392 m)
12	2	Glacier tongue	East	High	2021	Moderate (166 m)
1	3	Rock/ice face (shadow)	West	High	210	High (606 m)
2	3	Rock/ice face	North	High	815	High (617 m)
6	3	Rock/ice face (shadow)	South	High	553	High (537 m)
8	3	Rock/ice face (shadow)	South	High	347	High (723 m)
13	3	Rock/ice face	North	High	1567	High (652 m)

terrain type. For all methods, the RMSE values increase for regions of steeper slope.

The mean difference (ΔZ) between the three digital elevation models and the control data together with the RMSE for each AOI are presented in Table 3. The ΔZ were calculated by comparing each z-value of the digitized sampling point of the reference dataset against the cell value of the corresponding cell in the three ASTER derived DEMs. To ensure both the accuracy and precision of the methods are presented we provide both the mean ΔZ, to show any skew in the results in order to

Table 2. *Error statistics for the three terrain types*

Terrain type	Mean ΔZ (m)			RMSE (m)		
	NASA	SILCAST	ENVI	NASA	SILCAST	ENVI
1	−11.4	−14.2	−7.7	7.2	7.6	4.8
2	1.5	0.1	−4.8	10.8	8.7	17.0
3	−27.0	−41.0	45.1	197.0	195.7	158.1

Table 3. *Error statistics for the 13 sampling areas*

AOI no.	Terrain type	Mean ΔZ (m)			RMSE (m)		
		NASA	SILCAST	ENVI	NASA	SILCAST	ENVI
3	1	−16.0	−20.9	−9.4	7.1	3.3	3.8
4	1	−8.1	−10.0	−4.4	3.9	4.3	4.9
5	1	−7.0	−7.4	−6.9	4.0	4.3	4.0
9	1	−3.7	−3.3	−0.6	9.5	9.7	21.7
7	2	−7.3	−2.8	−13.4	9.8	10.0	34.9
10	2	6.7	0.4	−2.4	18.9	11.3	15.3
11	2	0.9	0.5	−20.2	15.8	14.4	25.9
12	2	1.9	0.4	−1.5	6.2	6.4	6.8
1	3	−253.0	−274.2	−17.8	140.8	138.1	77.5
2	3	−200.7	−213.8	−77.6	215.6	220.2	88.6
6	3	265.3	243.4	357.0	107.6	116.6	82.2
8	3	68.0	−16.9	130.6	36.4	51.3	28.1
13	3	−30.6	−25.5	−11.7	24.4	23.2	26.7

highlight systematic error, and the RMSE to show the spread of the results. Given the sample size involved in this case, we can consider RMSE to be equivalent to standard deviation.

There is a clear difference in the accuracy of ASTER DEMs between areas of low–moderate relief and areas of high relief. For terrain types 1 and 2 most DEMs are accurate to better than ± 30 m RMSE. This is in good agreement with the design specifications for the NASA relative DEM product of $\pm 10- \pm 30$ m RMSE (Fujisada *et al.* 2001; Hirano *et al.* 2003).

However, results are significantly worse for steeper ground of terrain type 3, where errors are often between ± 100 and ± 200 m RMSE. It is apparent that the accuracy of the DEMs is highly correlated to the steepness of the terrain. This result has been noted in other studies including Hirano *et al.* (2003) and Kääb *et al.* (2002, 2005). Overall, the error values obtained are very similar to those reported elsewhere for elevation data from ASTER data in mountainous terrain (Kääb *et al.* 2002, Kääb 2005). It is likely that slopes with high gradients will be more distorted and shadowed in the imagery, making image matching more difficult and resulting in increased errors. The relatively large size of the image pixels (15 m) may also mean that all peaks are not fully resolved in the imagery, resulting in a smoothing down of the resulting DEM.

For terrain types 1 and 2 there appears to be no clear best method of the three evaluated, although in some instances (AOIs 7, 9 and 11) ENVI performs slightly worse than the NASA–SILCAST approach. For steeper slopes of terrain type 3, where the errors are larger, it does appear that the ENVI DEM results are better than the NASA and SILCAST methods.

For each terrain type, histograms were generated showing the distribution of ΔZ (Figs 3–5). The *y*-axis count is the number of digitized 3D data points from the reference photogrammetric dataset within the AOIs of each terrain type. The distribution of errors for each terrain type is shown in Figures 3–5 and this allows comparison of each of the three DEM generation methods.

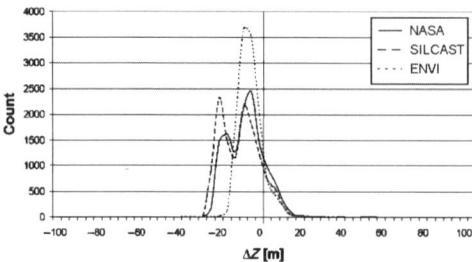

Fig. 4. Height error histogram for terrain type 2.

The peaks of all three histogram plots have negative ΔZ values. Therefore, we conclude that in most cases elevation values from ASTER satellite data are too low.

A notable feature of the histogram for terrain type 1 (Fig. 3) is the bimodal shape of the error distribution in the case of both the NASA and SILCAST methods. The errors from these methods for AOI 3 are larger than for the other AOIs of this terrain type. Owing to the relatively larger number of digitized reference points in this AOI, the histograms have a noticeable peak corresponding to these errors at *c.* $\Delta Z - 20$ m.

For terrain type 2, the histogram (Fig. 4) shows a noticeable negative bias to the ENVI results compared to the other two methods. This result shows that, in this terrain, ENVI mostly measures the DEM elevation too low, while the NASA and SILCAST methods calculate both negative and positive errors.

If the other characteristics (Table 1) of the sample regions are considered, there is little correlation between them and the resulting DEM accuracy results. There appears to be no correlation between the accuracy and surface texture. Given that the DEM generation methods rely on automatic image-matching methods, we would expect the correlation accuracy to be less in regions of low texture, leading to increased errors in the resulting elevation data. This would have a significant impact on the large areas of featureless snow and deep shadows encountered in the mountainous

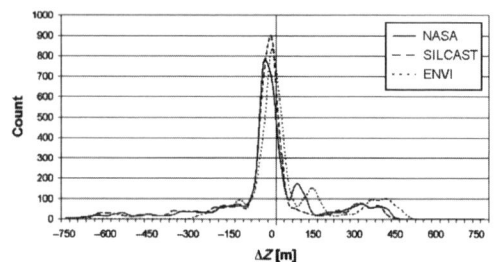

Fig. 3. Height error histogram for terrain type 1.

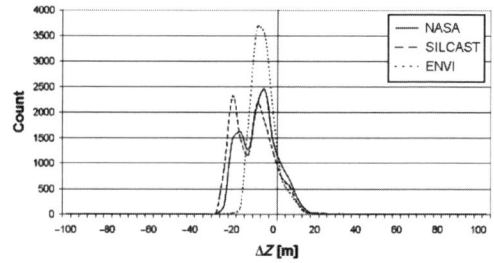

Fig. 5. Height error histogram for terrain type 3.

terrain being considered. Relative to the aerial photography, the larger 15 m pixel size of the ASTER satellite image 'smooths' a large amount of the natural surface detail present. In particular, crevasses on the glaciers are much more apparent in the higher resolution photography. Despite this, accuracy figures of better than 25 m are still attained in low relief areas that show very little surface texture.

Equally, there is no correlation between the DEM accuracy and slope aspect. Other studies (Kääb *et al.* 2002) have reported higher errors for regions with steep slopes facing away from the satellite sensor. They suggest that the rear-viewing ASTER band (3B) heavily distorts or obscures these slopes because of the 26.7° viewing angle, resulting in less accurate DEM values. However, this effect is not apparent in our results, and we cannot see any linkage between slope aspect and the quality of the DEM.

We know (M. Abrams pers. comm., JPL, NASA, January 2007) that the NASA and SILCAST approaches use the same method because NASA employs the SILCAST software to generate their product. However, the statistics show that the results are not identical. We expect that this is due to variations in the parameters used (e.g. output pixel size, correlation factors, numbers of tie points) and highlights the dangers of a 'black-box' approach where it is assumed the result is the best achievable for a particular region. It should not be assumed that one method is best in all instances.

Given the similar approaches used by NASA and SILCAST, for all sample areas the results from the NASA and SILCAST DEMs show a broadly similar pattern. This similarity is further reinforced when comparing contours generated from each of the three ASTER DEMs. Contours generated using Leica Photogrammetry Software for an example area in the west of the island from the NASA (Fig. 6) and SILCAST DEMs (Fig. 7) are markedly different in their shape and coverage when compared to contours generated from the ENVI DEM (Fig. 8).

The comparison of these contours highlights regions where the output of the three methods of DEM generation is relatively better or worse. The example in the figures again shows that the ENVI method is more effective for generating accurate elevation data for regions of steep and rocky ground than either the NASA or SILCAST method. This is highlighted for the rocky region shown SW of AOI 2 and for the area covered by AOI 1. For these regions the ENVI DEM provides realistic contours, and the errors for AOI 1 (Table 3) are better for the ENVI results when compared to the other two methods.

Some of the differences in performance between the ENVI methods and the NASA–SILCAST methods may be due to the different image-matching approaches used in the respective algorithms. Some studies (e.g. Brockelbank & Tam 1991; Kang *et al.* 1994) have suggested that natural surfaces tend to have too few structural features for

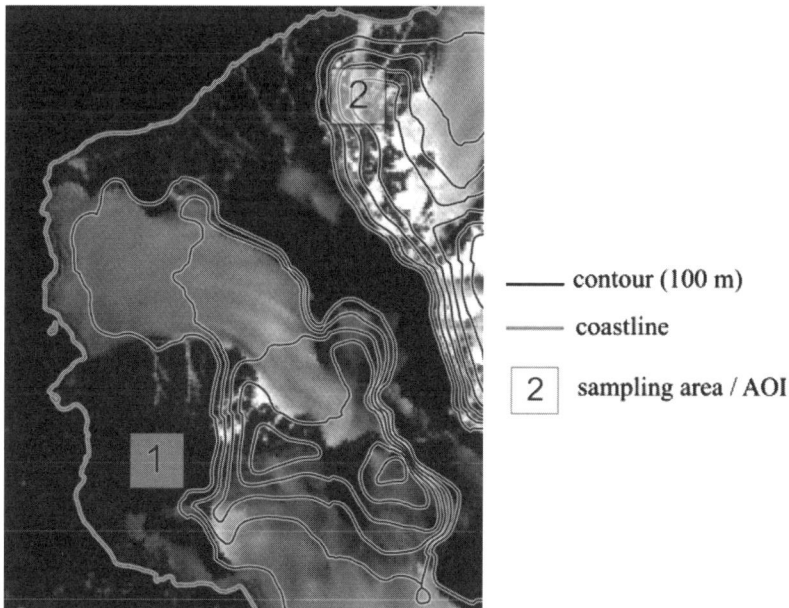

——— contour (100 m)

——— coastline

2   sampling area / AOI

Fig. 6. NASA contours.

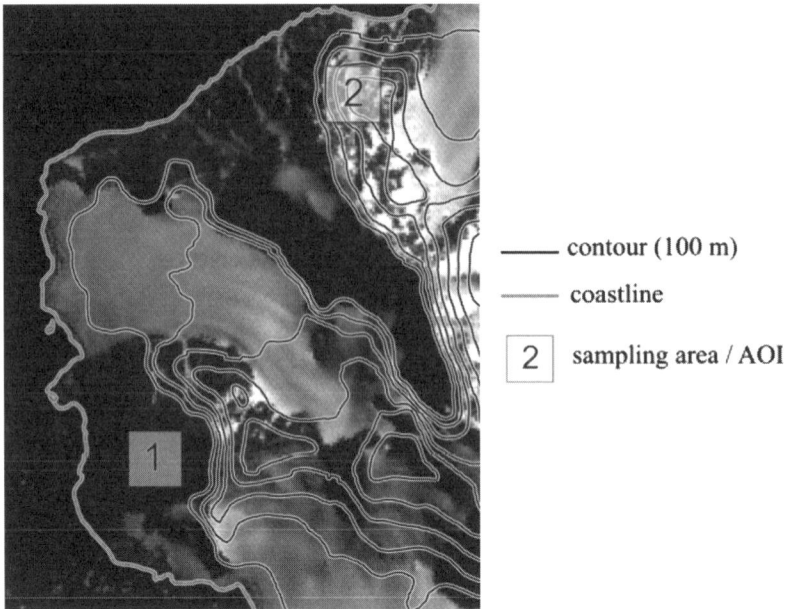

Fig. 7. SILCAST contours.

feature-based image matching to work well and that area-based methods are more effective.

The algorithm employed in the ENVI software uses a combined feature and area-based matching approach to image correlation (Shippert & Yang 2006). Documentation for the NASA algorithm (ERSDAC 2005) indicates that a purely area-based matching approach is used. While the correlation is not clear between the matching approach used and the results, it is possible that the inclusion of a

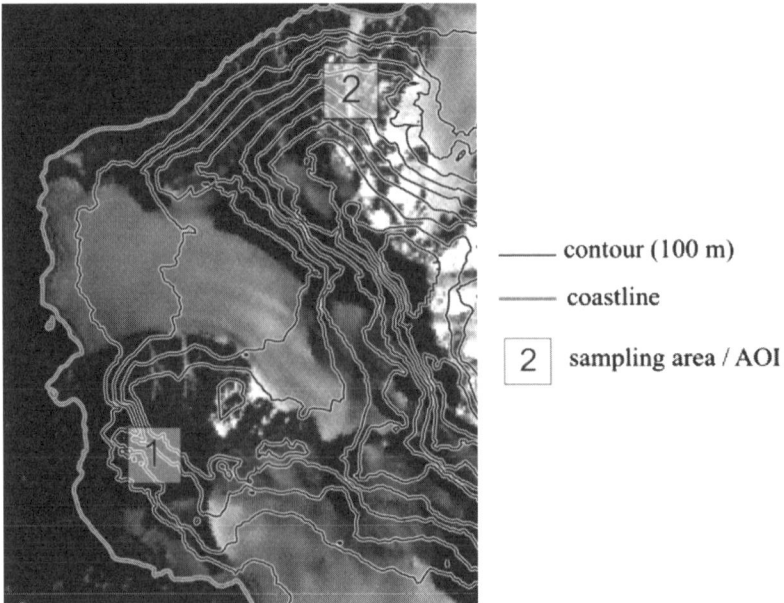

Fig. 8. ENVI contours.

feature-based approach in the ENVI method may improve results for some terrain types, while increasing the errors in others.

In regions of flat and smooth snow (e.g. AOI 9), where there is very little or no texture, an initial match based on features may introduce blunders that then propagate through a successive area-based matching. Feature-based matching approaches developed by Förstner & Gülch (1987) specify criteria for suitable feature points including the need to be distinct from neighbouring points, referred to as 'distinctiveness', and different on a regional scale to ensure separability across the whole image, referred to as 'seldomness'. Snow and glacier surfaces with locally repetitive features such as sastrugi (sharp irregular grooves or ridges formed on a snow surface by wind erosion) or surface melt-water streams could cause problems for the 'distinctness' criterion, and regionally repetitive features such as crevasse fields could cause problems with the 'seldomness' criterion, leading to matching failures or blunders. This may offer an explanation for the larger RMSE for AOI 9 for the ENVI result compared to the NASA–SILCAST results (Table 3). Conversely, in areas of low detail or heavy shadow but containing a few highly signalized features such as crevasses, where a purely area-based approach may have little useful data, an initial feature-based match may provide sufficient correlation between images to constrain the search area for the subsequent area-based matching, improving the results (e.g. AOI 1).

It is difficult to fully ascertain what factors affect the methods being used, but it is likely to be a complex interplay between the type and amount of surface texture and the filtering effect of the 15 m pixel size. Future work is planned to ascertain the relative roles of these factors in order to determine a more optimal method.

Conclusions

It is clear that the results are best in areas of low–moderate relief and performance is worse for steeper ground. There appears to be little or no correlation between the results and any other characteristics of the terrain types considered here, but it is highly likely that surface texture plays a key part in how well the methods work.

Of the methods considered, it does appear that ENVI performs better in steep terrain than the NASA and SILCAST approach. Results from all methods are very similar for regions of low–moderate relief.

The accuracy of elevation data from ASTER stereo imagery does not compare well to results from photogrammetry with aerial photography.

However, we suggest that for certain applications it offers elevation data of sufficient quality. Applications such as landform studies, 3D visualization, orthocorrection of satellite imagery and air photography, glacier extent mapping and glacial deformation studies (Kääb et al. 2002) can all make use of this data. For the majority of the Antarctic Peninsula it represents the best elevation data available and is frequently an order of magnitude better in resolution than any other DEM available. Other factors such as the low cost per unit area and its independence from ground-based measurements make it especially attractive in remote and inaccessible polar and mountainous regions.

References

AL-ROUSAN, N. & PETRIE, G. 1998. System calibration, geometric accuracy testing and validation of DEM and orthoimage data extracted from SPOT stereo pairs using commercially available image processing systems. *International Archives of Photogrammetry & Remote Sensing*, **32**, 8–15.

BROCKELBANK, D. C. & TAM, A. 1991. Stereo elevation determination technique for SPOT imagery. *Photogrammetric Engineering and Remote Sensing*, **57**, 1065–1073.

CHENG, X., ZHANG, Y., DONGCHEN, E., LI, Z. & SHAO, Y. 2003. Digital elevation model construction using ASTER stereo VNIR scene in Antarctic in-land ice sheet. *In: Geoscience and Remote Sensing Symposium, 2003. IGARSS '03. Proceedings 2003 IEEE International*, **5**, 3347–3349.

EDC DAAC 2001. ASTER DEM data product. Available online at: http://edcdaac.usgs.gov/aster/ast14dem.asp.

ERSDAC 2005. *ASTER User Guide Part 3, DEM Product (L4A01) Version 1.1*. Earth Remote Sensing Data Analysis Centre, NASA. Available online at: http://www.science.aster.ersdac.or.jp/en/documnts/users_guide/index3D.html.

FARR, T. G., HENSLEY, S., RODRIGUEZ, E., MARTIN, J. & KOBRICK, M. 2000. The Shuttle Radar Topography Mission. *In: Proceedings of the CEOS SAR Workshop, 26–29 October 1999, CNES, Toulouse*. ESA-SP, **450**, 361–363.

FÖRSTNER, W. & GÜLCH, E. 1987. A fast operator for detection and precise location of distinct points, corners and centers of circular features. *In: Proceedings of the Intercommission Conference on Fast Processing of Photogrammetric Data, Interlaken, 1987*. International Society for Photogrammetry and Remote Sensing (ISPRS), Beijing, 281–305.

FUJISADA, H., IWASAKI, A. & HARA, S. 2001. ASTER stereo system performance. *In*: FUJISADA, H., LURIE, J. B. & WEBER, K. (eds) *Sensors, Systems, and Next-Generation Satellites V. Proceedings of SPIE*, **4540**, 39–49.

HIRANO, A., WELCH, R. & LANG, H. 2003. Mapping from ASTER stereo image data: DEM validation and accuracy assessment. *ISPRS Journal of Photogrammetry and Remote Sensing*, **57**, 356–370.

KÄÄB, A. 2005. Combination of SRTM3 and repeat ASTER data for deriving alpine glacier flow velocities in the Bhutan Himalaya. *Remote Sensing of the Environment*, **94**, 463–474.

KÄÄB, A., HUGGEL, C., PAUL, F., WESSELS, R., RAUP, B., KIEFFER, H. & KARGEL, J. 2002. Glacier monitoring from ASTER Imagery: accuracy and applications. *Proceedings of EARSeL–LISSIG–Workshop Observing our Cryosphere from Space, Bern, 11–13 March 2002. EARSeL eProceedings*, **2**, 43–53.

KÄÄB, A., HUGGEL, C. *ET AL.* 2005. Glacier hazard assessment in mountains using satellite optical data. *EARSeL eProceedings*, **4**, 1/2005, 79–93.

KANG, M. S., PARK, R. H. & LEE, K. H. 1994. Recovering an elevation map by stereo-modelling of the aerial image sequence. *Optical Engineering*, **33**, 3793–3802.

LANG, H. & WELCH, R. 1999. *NASA ASTER DEM Algorithm Theoretical Basis Document, Version 3.* Available online at: http://eospso.gsfc.nasa.gov/eos_homepage/for_scientists/atbd/docs/ASTER/atbd-ast-14.pdf.

REES, W. G. 2005. *Remote Sensing of Snow and Ice*. CRC Press, Boca Raton, FL.

SHIPPERT, P. & YANG, Z. 2006. Extracting DEMs from Stereo Imagery. *Geoconnexion International Magazine*, February, 26–27. Available online at: www.geoconnexion.com.

Filtering of ASTER digital elevation models using mathematical morphology

G. DANNEELS*, H. B. HAVENITH, F. CACERES, S. OUTAL & E. PIRARD

Université de Liège, Place du 20-Août, 7, 4000 Liège, Belgium

Corresponding author (e-mail: gaelle.danneels@gmail.com)

Abstract: This paper presents results of research undertaken on the creation and filtering of digital elevation models (DEMs) from a stereo pair of Advanced Spaceborne Thermal Emission and Reflection Radiometer (ASTER) images. The raw DEM, created by automatic image matching, appears to be very noisy. Two types of irregularities can be observed. First, a random occurrence of small sinks and mounds with high amplitude is observed. Secondly, a more regular east–west-oriented pattern of noise is present. Many DEM-creation programs provide some editing tools to smooth out the irregularities, including some noise removal, smoothing and interpolation algorithms. However, the application of these algorithms has an important impact on the values of the parameters derived from the elevation, such as slope, aspect and curvature. In this study we propose a filtering algorithm based on morphological greyscale reconstruction in order to remove the sinks and mounds. This technique is very effective in mitigating the artefacts while preserving the remaining structures. For the regular pattern, a linear north–south-oriented low-pass filtering showed the best results. This approach was compared with a median filter and proved to be more effective in terms of both elevation and slope parameters.

Digital elevation models (DEMs) have been increasingly used for terrain monitoring and remote sensing applications. Moreover, elevation-derived parameters (slope, aspect and curvature) are often used as terrain variables for further analyses such as morphological analyses and drainage network investigations. However, artefacts, that is, artificial errors generated during the creation process of the DEM, often affect the DEMs. There is thus a need for editing in order to improve the accuracy of the elevation values. The filtering step is especially important for the terrain modelling applications, since the errors will usually propagate and be enhanced in the elevation-derived variables (Bolstad & Stowe 1994; Woods 1996).

Nowadays, various techniques have been developed for the acquisition of elevation data, characterized by different resolutions and accuracies. A widely used technique for the generation of DEMs is digital photogrammetry, where elevation parallaxes are extracted by image matching of a pair of images with stereoscopic view. The spatial resolution of the created DEM is directly related to the spatial resolution of the stereoscopic images (aerial photographs or satellite images). The height accuracy is dependent upon the ground pixel resolution, the base-to-height ratio of the acquisition system and the reliability of the correlation procedure (Kasser & Egels 2001). For DEMs created with ASTER images, the height accuracy can be roughly estimated to be between 12 and 25 m.

An important and well-known problem using this technique is the presence of artificial 'sinks' and 'mounds' produced for regions with low spectral contrast and thus poor correlation between the images (Toutin 2002; Hirano *et al.* 2003; Gamache 2004; Stevens *et al.* 2004; Fujisada *et al.* 2005). Therefore, there is always a need for post-processing the extracted elevation model. As manual refinement is very time-consuming, most DEM generation software provides tools to remove the irregularities and create a more pleasing DEM. They usually include some noise removal functions based on statistical parameters (mean or standard deviation) and interpolation algorithms to replace the mismatched values. In addition, they propose filter algorithms such as the moving mean and median filtering (PCI 2003; ENVI 2005). However, their ability to reduce noise is accompanied by a loss of image (or DEM) detail, and thus smoothing of the image. This smoothing effect has an important impact on the further analysis of DEMs, as it affects the values of elevation-derived parameters.

The simplest way to reduce the number of these mismatching errors is to create DEMs at lower resolution. As most end-users of the ASTER data will use the DEMs for regional analyses, which do not require very high resolutions, this is the most common option (see review in Gamache 2004). As a result, very little research has been undertaken to develop adequate filters for the artefacts. Wang (1998) introduced a two-dimensional (2D) Kalman filter to produce optimal estimates of elevation-derived variables from a noisy DEM. His approach consists of a 2D Kalman processor, a function for

From: FLEMING, C., MARSH, S. H. & GILES, J. R. A. (eds) *Elevation Models for Geoscience.*
Geological Society, London, Special Publications, **345**, 33–42.
DOI: 10.1144/SP345.5   0305-8719/10/$15.00 © The Geological Society of London 2010.

the detection and removal of outliers, and a two-step filtering procedure. Trinder *et al.* (2002) compared this Kalman filter with a filter method based on wavelet transforms, more specifically the *à trous* (with holes) algorithm. Some filtering methods for more regular processing artefacts can be found. Oimoen (2000) proposed a mean profile algorithm for removing striping discontinuities in USGS 7.5-min DEMs. Albani & Klinkenberg (2003) describe a line-based cross-smoothing algorithm for the removal of striping artefacts of DEMs.

Unlike DEMs produced by photogrammetry, a lot of research has been undertaken for the editing process of LIDAR (Airborne Light Detection and Ranging) data. The major concern with this technique is to discriminate the ground points from the off-terrain points (such as trees, buildings and vehicles). The non-ground points can be removed using mathematical morphology filters. This technique requires the irregularly spaced LIDAR points to be converted into a regular greyscale grid image. Kilian *et al.* (1996) proposed some early ideas to use morphological opening operations to discriminate the ground points. As there is no optimal size for the structuring element, they suggest the use of a series of openings with different structuring element sizes. Zhang *et al.* (2003) developed a progressive morphological filter in order to enable automatic extraction of the ground points where the algorithm uses the classical morphological opening with a gradually increasing size of the structuring element in combination with elevation difference thresholds. Arefi & Hahn (2005) presented a new method based on morphological greyscale reconstruction where the marker image is generated by subtracting a constant value from the mask image; thus, there is no need to specify a kernel size. All maxima whose depth are smaller than the given threshold will be suppressed.

In this paper the artefacts of DEMs created by digital photogrammetry from a pair of ASTER images are analysed. Regarding the sinks and mounds, they could be seen as off-terrain points. Therefore, a filtering technique using morphological greyscale reconstruction combined with a linear low-pass filtering is proposed. This technique has been tested in a mountainous region strongly affected by artefacts. The filtered DEM was compared with data derived from a topographical map with scale 1:25 000.

Mathematical morphology filters

Mathematical morphology is a powerful image analysis technique making use of non-linear processing algorithms. The morphological operators can be defined as neighbourhood image operators, that is, the output value of an image pixel is obtained by a combination of the image values lying within its neighbourhood (Soille 1999). The structure and size of the neighbourhood are defined by elementary geometrical forms called structuring elements (SE) and are usually chosen according to *a priori* knowledge of the geometry of the features to be extracted. In this paper we will follow the notations used by Soille (1999). All morphological operators can be applied to binary images or greyscale images. In morphology, every greyscale image is regarded as a topographical surface with the grey-level intensities corresponding to elevation values. The morphological operators can then be defined in terms of minimum and/or maximum of the neighbouring 'elevation' values. This technique is thus suited for the analysis of DEMs with a regular grid configuration.

The two basic morphological operators are erosion and dilation. The erosion of a greyscale image, f, by a structuring element, B, is denoted by $\varepsilon_B(f)$. The eroded value of a given pixel is defined by the minimum value of the image within the neighbourhood defined by the SE. The dual operator of dilation of an image, f, is denoted by $\delta_B(f)$ and can be defined as the maximum value of the image in the window defined by the SE. By combination of the two basic operators (erosion and dilation), new morphological transformations can be generated. The morphological opening of f by B is defined as the erosion of the image f by the SE B, followed by a dilation with the transposed SE $\check{B}$ (Soille 1999):

$$\gamma_B(f) = \delta_{\check{B}}[\varepsilon_B(f)].$$

Applying this operation to a greyscale image will erase all peaks that are smaller than the SE, and thus filter out the bright image structures (see Fig. 1b). The closing operation, denoted by $\phi_B(f)$, is accomplished by performing dilation followed by erosion:

$$\phi_B(f) = \varepsilon_{\check{B}}[\delta_B(f)].$$

The application of closing to a greyscale image can be seen as a filling up of the sinks smaller than the SE. It will thus filter out all dark image structures smaller than the SE.

However, these processes of classical greyscale morphology also alter the shape of the components that are not removed. This disadvantage does not occur with morphological reconstruction, which is based on the concept of 'geodesic' propagation (Vincent 1993). Morphological reconstruction is based on the manipulation of two input images (a mask image and a marker image), instead of one input image and a SE for classical morphology. The elementary geodesic dilation is defined as the

Fig. 1. Example of the classical opening and opening by reconstruction on a greyscale image with a square SE B of size 12×12 pixels. (**a**) Original image f. (**b**) Classical opening of f: $\gamma_B(f) = \delta_B[\varepsilon_B(f)]$. (**c**) Opening by reconstruction of f: $\gamma_R(f) = R_f[\varepsilon_B(f)]$.

point-wise minimum between the mask image, g, and the elementary dilation of the marker image, f (Soille 1999). In other words, the marker image, f, is first dilated by the elementary SE and the resulting dilated image is then forced to remain below the mask image, g.

The greyscale reconstruction of mask image, g, from a marker image, f, is obtained by iterating greyscale geodesic dilations of f 'under' g until stability is reached (Vincent 1993). This process can be used for performing an opening by reconstruction. The original image, f, is first eroded by a specific SE of size n, as for the morphological opening. The eroded output is then used as the marker image for the reconstruction, while the original image is used as mask image and controls the expansion of the marker image:

$$\gamma_R^{(n)}(f) = R_f[\varepsilon^{(n)}(f)].$$

By applying this filter all features (mounds or sinks) that are smaller than the SE are filtered out, while the remaining features are restored to their original shape. The result of an opening by reconstruction is illustrated in Figure 1c.

Furthermore, the opening and closing operations mentioned above (both classical and by reconstruction) are characterized by the following properties, which cannot be defined for the median and linear filters.

- The opening and closing are *idempotent*; that is, applying the transformation twice to an image is equivalent to applying it only once. Therefore, whereas for the median and linear filters the final result is highly dependent on the number of times the filter is applied, the idempotence property makes an arbitrary choice of the number of iterations unnecessary.
- The closings are *extensive*; that is, the transformed image is greater than or equal to the original image, while the openings are *anti-extensive*, that is the transformed image is smaller than or equal to the original. We can thus choose to

specifically erase peaks (or bright image structures) with the opening while leaving the holes and valleys unaltered, whereas the moving mean or median filter will alter both simultaneously.

These properties are very useful, as they will allow us to predict the transformed result and thus help us to choose the appropriate transformation.

DEM creation and artefact characterization

For the present study, DEMs were created on the basis of pairs of stereoscopic Level 1B ASTER images, using the 3N and 3B bands in along-track configuration. The ASTER scenes are processed using digital photogrammetric techniques: the elevation parallaxes are extracted by automatic image matching of the two scenes using a multiscale mean-normalized cross-correlation method (Toutin 2004). Measured parallax values are then converted to relative elevations using trigonometry and the satellite orbital data. An absolute geocoded DEM was created by adding 15 user-specified ground control points (GCPs), evenly distributed in the scene in both planimetry and elevation.

Pixels for which the parallax (and thus elevation) values are calculated are defined by a user-fixed sampling interval. The elevation values for the pixels in between are inferred afterwards by interpolation. A small interval allows rendering of all the small-scale features, but also results in more matching errors. Using a larger interval results in fewer matching errors, but this will already induce a smoothening of the data. For example, when choosing a sampling interval of 2 the parallax values are calculated for every second pixel. This is equivalent to deleting the elevation information of every second pixel, and replacing it with an average of the surrounding pixel values (weighted or not depending on the interpolation method). This will be useful in the case of very small artefacts (smaller than the chosen sampling interval) that are equally

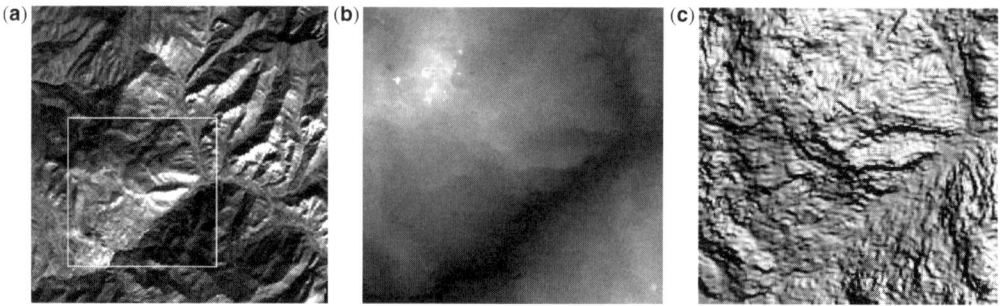

Fig. 2. First test area located in the Kyrgyz Tien Shan Range. (**a**) 3N band. (**b**) Greyscale image of DEM. (**c**) Shaded relief view of DEM. (a) Subzone (400 × 400 pixels) of the 3N band. Zoom inside white rectangle with visualization of the mounds (white spots) and sinks (dark spots) in (b) and visualization of the east–west-oriented 'wavy' pattern in (c).

distributed all over the DEM (e.g. salt and pepper noise). However, in case of very local and larger errors, using a larger interval will smooth them out but not remove them completely. As it is easier to filter out some well-defined structures, we decided to work with the highest detail (i.e. parallax calculation at every pixel) and a resolution of 15 m (same as the original images).

A DEM was created with images acquired during the summer of 2001, located in the Kyrgyz Tien Shan Range. For further quality assessments, we restricted the area to a small subzone of 400 × 400 pixels (6 × 6 km), characterized by a mountainous topography (elevations between 1500 and 2500 m) with light vegetation (see Fig. 2a).

Two kinds of artefacts can be observed. First, the well-known 'sinks' and 'mounds' (see the start of this paper) are present, which are produced by matching errors between the images. These artefacts have a size of up to 5 × 5 pixels and cause elevation errors up to 100 m. They are visible as small bright and dark spots in Figure 2b. Second, we observe a more regular pattern of 'wavy' structures, with a wavelength up to 7 pixels and a magnitude up to

25 m. These irregularities appear to be correlated with an abrupt lowering of the intensity values in the stereo images, for example, at the boundaries of a clearly delimited river or at the limit between the exposed and shaded sides of mountains. In Figure 2c the sun is shining from the south (solar azimuth angle of 163°), and abrupt changes occur between north- and south-facing flanks. Maxima and minima of the waves thus mostly follow east–west-oriented patterns. These kinds of artefacts were also observed on a DEM of the same zone, created from ASTER images acquired during the summer of 2004 (solar azimuth angle of 146°).

No reference to these kinds of artefacts could be found in the literature. This is probably due to the fact that these artefacts are smoothed out during the DEM creation process when using a lower parallax sampling interval. Therefore, a second DEM was created in another area to check the presence of these 'wavy' structures. This second test area is located in the Bolivian Altiplano, and is characterized by an arid climate with very little vegetation and a more gentle topography (elevations between 4500 and 5000 m; see Fig. 3a). The same

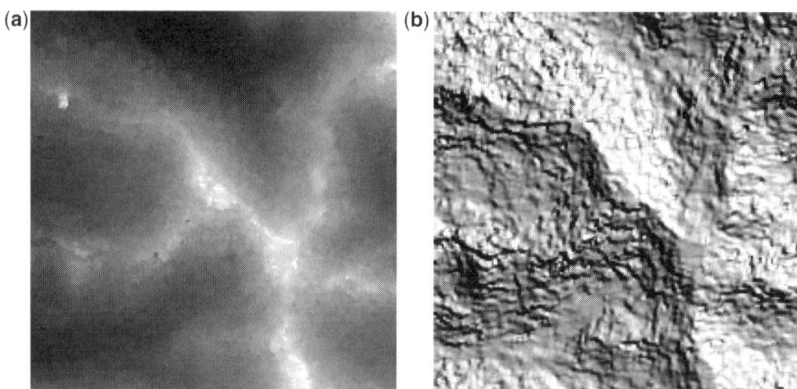

Fig. 3. Second test area located in the Bolivian Altiplano. (**a**) Greyscale image of DEM. (**b**) Shaded relief view of DEM. (a) Subzone (200 × 200 pixels) of the DEM. (b) Visualization of the NW–SE-oriented 'wavy' pattern.

artefacts are also present in this second test area, where the 'wavy' structures are now mainly NW–SE oriented (solar azimuth angle 93°), following the main topographical structures (see Fig. 3b).

DEM filtering

In this section we will examine both classical moving average and median filtering, as well as a filter based on mathematical morphology. The most common and widely used noise removal filter is the linear low-pass filter, also known as the moving average filter. This filter replaces each pixel value with the average of the surrounding values, thus preserving only the low-frequency components of the image. However, this yields a general smoothing. In this respect, the moving median filter performs better than the mean filter, as it is able to preserve sharp edges. Our aim is to preserve as much of the original information of the DEM as possible; that is, to keep smoothing as low as possible. We are thus looking for a filter that will remove the artefacts while producing the best estimates both in terms of elevation and elevation-derived parameters. This is made possible by choosing filters that are best adapted to the artefacts, with regard to the method as well as the shape and size of the structuring element. In our case we are dealing with two types of artefacts with different characteristics; therefore, we propose an approach with two different filters that will selectively remove one type of artefact. Furthermore, as it is easier to remove distinct features, it is particularly important to be able to eliminate one kind of artefact without altering the other kind.

Filtering was tested on a small subzone (200 × 200 pixels) within the Kyrgyz Tien Shan Range, where the resulting DEMs could be compared with a DEM derived from a 1973 topographical map with a 1:25 000 scale, which will be used as a reference DEM.

Filtering of mounds and sinks

In our case the first kinds of artefacts are 'sinks' and 'mounds', which are very local errors requiring a specific filter. Application of an opening will filter out the bright image structures (i.e. the mounds), while the closing will have the same filtering effect on the dark image structures (i.e. sinks). As mounds as well as sinks corrupt the DEM, it is necessary to use a sequential combination of an opening and closing. The order of the combination, that is, opening followed by closing or vice versa, produces almost the same result but they are not equivalent (Soille 1999). An open–close filter (opening followed by closing) will enhance the darker structures, whereas a close–open filter will enhance the brighter structures. A solution to lower the impact of combination order is to use an alternating sequential filter (ASF). This is a sequential application of closings and openings, beginning with a small SE and increasing the size of the SE at each step until a given size n:

$$ASF_n(f) = \gamma^{(n)}[\phi^{(n)} \ldots (\gamma^{(2)}[\phi^{(2)}(\gamma^{(1)}[\phi^{(1)}(f)])])]$$

The shape and size of the structuring element has to be chosen as a function of the artefact to be removed. In this case, we will use an ASF starting with a closing, and composed of two sequences with SEs of 3 × 3 and 5 × 5 pixels, in accordance to the size of the mounds and sinks:

$$ASF_{(5 \times 5)}(DEM) =$$
$$\gamma^{(5 \times 5)}[\phi^{(5 \times 5)}(\gamma^{(3 \times 3)}[\phi^{(3 \times 3)}(DEM)])]$$

Indeed, the morphological transformations are based on the concepts of minimum and maximum: in order to change the value of a pixel inside an artefact, there only needs to be one of the neighbourhood pixels (defined by the SE) outside the artefact.

Figure 4a shows an area of the raw DEM affected by a mound (white zone) and a sink (black zone). The result of the ASF with classical greyscale closings and openings is shown in Figure 4b. We can see that this ASF manages to remove the mound and sink, but it also affects the entire DEM by creating some kind of platform. Thus, we propose to apply the ASF with openings and closings by reconstruction ($ASF_{(R)}^{(5 \times 5)}(DEM)$), as only the artefacts will be filtered and the rest will be kept unmodified. This technique has been applied by Outal (2006) on images of fragmented rocks in order to filter out small structures on the fragment faces, while preserving the contours of the fragments. Figure 4c shows the result of the ASF with reconstruction. This filter succeeds in removing the sink and mound, while leaving the rest unaltered.

These results are compared with a median filtering of 5 × 5 pixels in Figure 4d. We can clearly see that the median filtering has not filtered out the artefacts (see the profile in Fig. 4e) and induces some smoothing of the background. In order to manage to filter out the artefacts completely with the median filter a larger SE should be used. Indeed, this filter is based on the median of the neighbourhood values defined by the kernel. Thus, in order to change the value of a pixel inside the artefact, at least half of the kernel should be outside the artefact. However, using a larger SE will also yield more smoothing of the background.

The proposed ASF of openings and closings with reconstruction presents the best results. Unfortunately, this reconstruction principle has drawbacks when an artefact is connected to a larger structure

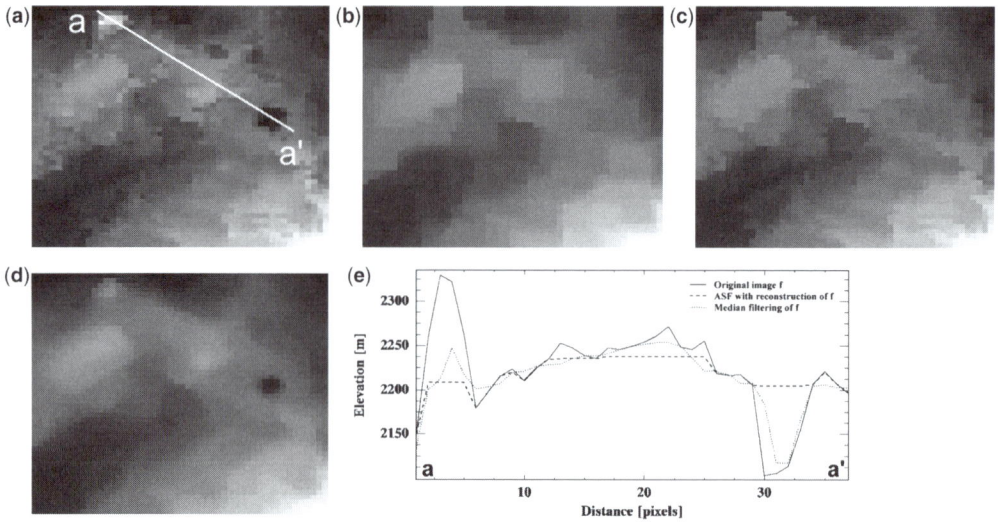

Fig. 4. Visualization of filtering of 'mounds' and 'sinks' in a subzone (35 × 50 pixels) of the Tien Shan test area.
(**a**) Original image *f*. (**b**) Classical ASF of *f*: $ASF_{(5\times5)}(f)$. (**c**) ASF with reconstruction of *f*: $ASF_R^{(5\times5)}(f)$. (**d**) Median Filtering of *f*: $Median_{(5\times5)}(f)$. (**e**) Elevation profile a–a′. (**a**) Greyscale image of a raw DEM with the location of the profile line. (**b**)–(**d**) Greyscale image of filtered DEMs. (**e**) Elevation profile a–a′ showing a comparison of filtering results.

of the same type. For instance, in the case of a mound connected to a larger ridge, the erosion process will successfully remove the mound. But, owing to the connectivity with the larger ridge, the mound will be (partially) reconstructed during the geodesic dilation operation. The same problem will occur for a sink connected to a larger valley.

Quantitative results of the subtraction between the topographical reference DEM and the raw and filtered DEM, respectively, are the following (negative errors caused by mounds; positive errors by sinks):

	Mean	Min	Max
Topographical – raw DEM	−8	−193	169
Topographical – filtered DEM	−8	−132	81

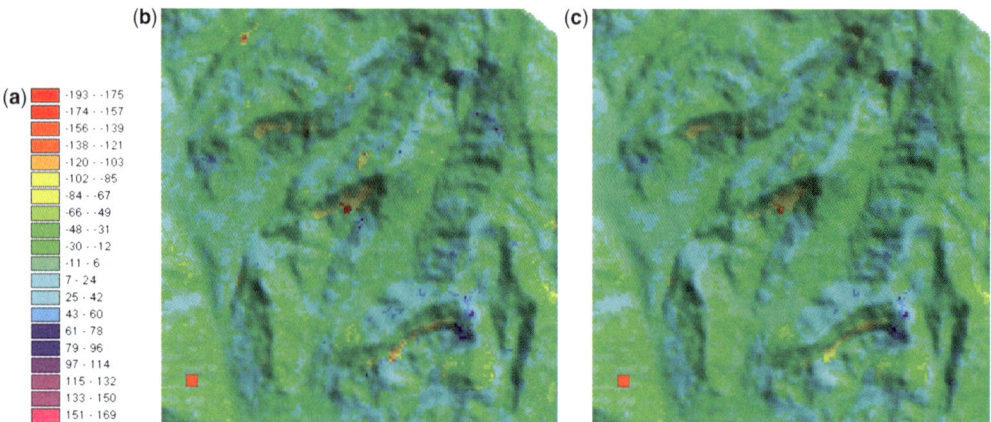

Fig. 5. Evaluation of the filtering results of 'mounds' and 'sinks'. (**a**) Elevation colour bar. (**b**) Topo – raw DEM. (**c**) Topo – filtered DEM. (**b**) & (**c**) Difference maps between reference DEM and raw and filtered DEMs, respectively, superposed on the shaded relief view of the reference DEM. The SE of size 5 × 5 pixels is represented by the red square in the lower-left corner.

Although very large errors are still present, the lowering of absolute values of minimum and maximum differences show the improvements of the filtered DEM compared to the raw DEM. In Figure 5 the difference maps are represented, and we can clearly see that almost all artefacts smaller than the SE (red square in the lower-left corner) have been removed.

Most of the remaining errors are caused by inaccuracies during the georeferencing process and/or by the poor quality of the topographical map, and thus do not reflect artefact errors that can be removed by filtering. They are typically found on the flanks of the hills and have an elongated shape. Other remaining errors are of the same type as the mound and sink artefacts, but with larger planimetric extent than the chosen SE. They are thus not recognized as artefacts and will not be removed with the proposed filtering technique.

Filtering of the wavy structures

The second type of artefact, the wavy pattern, does not have a large influence on the elevation values (error values are within the vertical accuracy of the DEM). However, it creates very large errors in the elevation-derived parameters such as the slope. In the first test area (Tien Shan Range), the east–west orientation of this pattern induces a small-scale

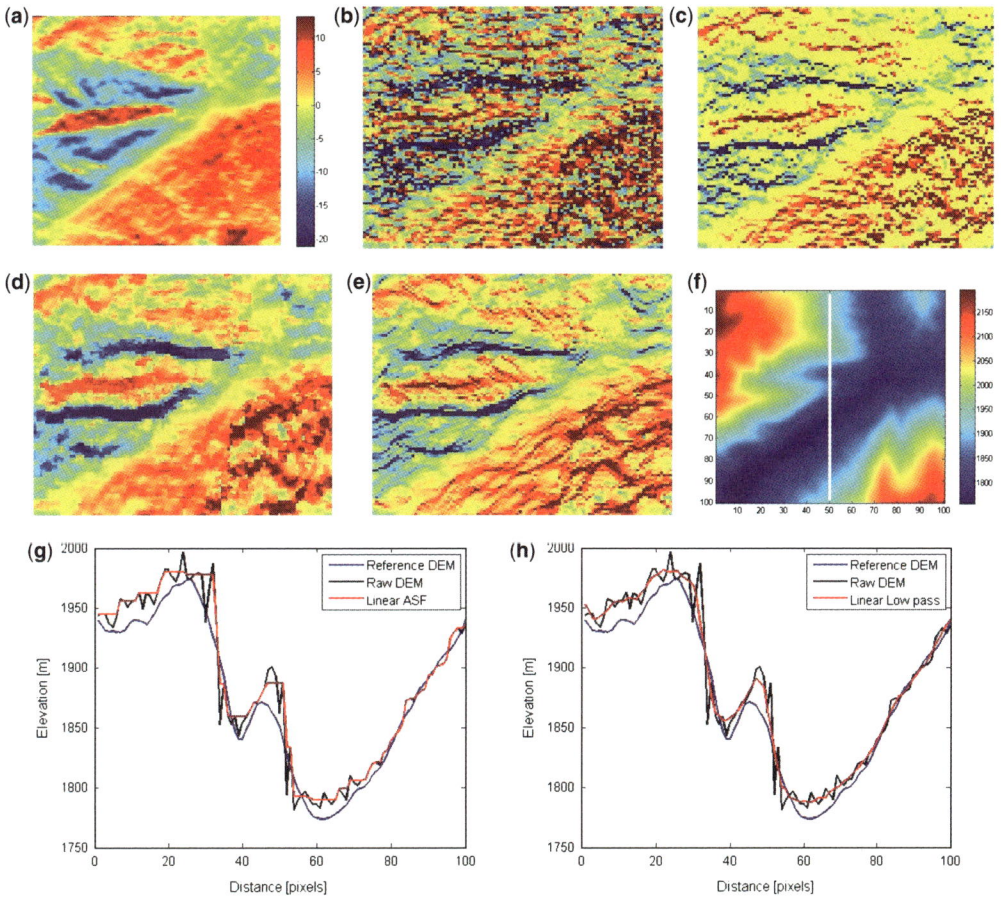

Fig. 6. Visualization of the filtering of the 'wavy' pattern. (**a**) Slopes of reference DEM. (**b**) Slopes of raw DEM. (**c**) Slopes after linear ASF of DEM: $ASF_{(5\times1)}(DEM)$. (**d**) Slopes after linear low pass filtering of DEM: $LP_{(5\times1)}(DEM)$. (**e**) Slopes after median filtering of DEM: $Median_{(5\times5)}(DEM)$. (**f**) Corresponding reference DEM. Comparison of slope maps in the north–south direction derived from (a) the topographical reference DEM, (b) the raw DEM and (c)–(e) the filtered DEMs. The slope is defined as the elevation difference between two neighbouring pixels in the north–south direction, with positive values for slopes towards the north and negative values for slopes towards the south. (f) Corresponding elevations of the reference DEM with an indication of the profile line. (**g**) & (**h**) Elevation profiles for the comparison of filtering results.

(a)

(b)

(c)

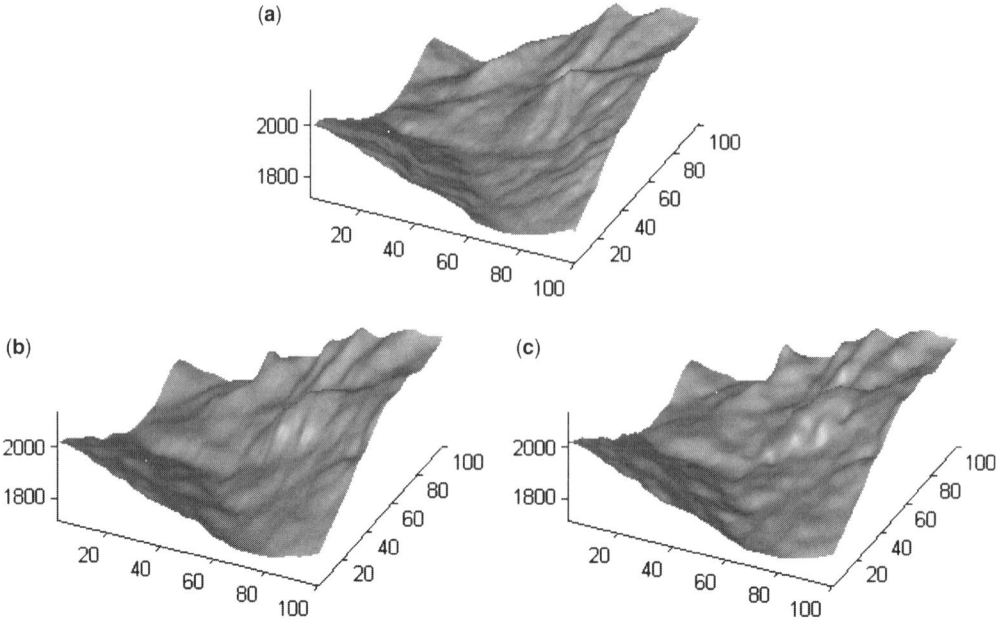

Fig. 7. Perspective view of (a) the reference DEM and (b) & (c) the filtered DEMs. (**a**) Reference DEM. (**b**) Reconstruction and low pass filtering. (**c**) Median filtering.

variability of the slope in the north–south direction. This can be clearly seen in Figure 6b, where the slope is defined as the elevation difference between two neighbouring pixels in the north–south direction.

A north–south-oriented elevation profile of the raw DEM is illustrated in Figure 6g, where we can see the artificial wave-like structures superimposed on the mountain flanks. As the artefacts are mostly east–west oriented, we suggest the use of a kernel of 5 × 1 pixels in a north–south direction, according to the average size of the 'waves' (slightly larger than half the length of a wave). As for the previous case, we could think of using a closing for filling up the lower half of the wave and an opening for the upper half. Figure 6c shows the resulting slope map after application of an ASF of classical openings and closings with kernel sizes 3 × 1 and 5 × 1 on the DEM:

$$ASF_{(5 \times 1)}(DEM) =$$

$$\gamma^{(5 \times 1)}[\phi^{(5 \times 1)}(\gamma^{(3 \times 1)}[\phi^{(3 \times 1)}(DEM)])]$$

We can see that the application of mathematical morphology transformations is not appropriate for this kind of artefact. As these transformations are based on the minimum or maximum of a flat horizontal SE, they will induce some artificial terracing effect on the flanks (see the profile in Fig. 6g).

We will thus have to use an averaging filter for removing these artefacts. The resulting slopes after applying a linear low-pass (LE) filter with kernel 5 × 1 on the DEM ($LP_{(5 \times 1)}(DEM)$) are shown in Figure 6d. The north–south profile of the resulting DEM (Fig. 6h) shows that the waves could be suppressed, and that the shape approaches the shape of the reference DEM. When comparing the result with the 5 × 5 median filter ($Median_{(5 \times 5)}(DEM)$, see Fig. 6e), we can see that the linear low-pass filter manages to filter out the waves more effectively, creating a slope map that is more similar to the topographical reference slope map.

Summary

Digital elevation models produced by automatic image matching of ASTER scenes include a lot of noise. The most commonly used solution is to create a DEM at lower resolution, followed by an additional editing of the DEM with the available noise removal, smoothing and interpolation algorithms. However, this procedure is accompanied by a loss of detail, and has an important impact on the elevation-derived parameters. Therefore, we propose to create a DEM at the highest possible resolution followed by a filtering that is more adapted to the artefacts and which induces less smoothing.

A first kind of artefact, due to mismatching during the correlation process, appears as random mounds and sinks. An alternating sequential filter of closings and openings by reconstruction has shown to be very effective in removing these artefacts and leaving the rest of the DEM unaltered. A second kind of irregularity, correlated to boundaries with high-intensity contrasts, appears as a more regular pattern of east–west-oriented 'wavy' structures. These artefacts could be removed by using a linear north–south-oriented low-pass filter.

The result of this combination of filtering techniques has been compared with a classic median filter. A 3D view of the final outputs is given in Figure 7. By visual comparison, we can see that the texture of the surface is better rendered with the newly proposed approach than with the median filter. The proposed approach is more effective for the removal of the artefacts, and gives more reliable results both in terms of elevation and elevation-derived parameters. Nonetheless, one should be aware that the proposed filtering technique does not take into account any hydrological concepts. Additional editing of the DEM (Hutchinson 1989; Soille et al. 2003; Soille 2004) might be necessary if the DEM is to be used for hydrological applications such as water-basin delineation or steepest slope path calculation.

References

ALBANI, M. & KLINKENBERG, B. 2003. A spatial filter for the removal of striping artifacts in digital elevation models. *Photogrammetric Engineering & Remote Sensing*, **69**, 755–765.

AREFI, H. & HAHN, M. 2005. A morphological reconstruction algorithm for separating off-terrain points from terrain points in laser scanning data. *In*: *ISPRS Workshop 'Laser Scanning 2005', Enschede, the Netherlands, September 12–14*. International Society for Photogrammetry and Remote Sensing (ISPRS), Beijing.

BOLSTAD, P. V. & STOWE, T. 1994. An evaluation of DEM accuracy: elevation, slope, and aspect. *Photogrammetric Engineering & Remote Sensing*, **60**, 1327–1332.

ENVI 2005. *ENVI Version 4.2 – User Manual*. RSI ENVI, Boulder, CO.

FUJISADA, H., BAILEY, G. B., KELLY, G. G., HARA, S. & ABRAMS, M. J. 2005. ASTER DEM performance. *IEEE Transactions on Geoscience and Remote Sensing*, **43**, 2707–2714.

GAMACHE, M. 2004. Free and low cost datasets for international mountain cartography. *In*: *4th ICA Mountain Cartography Workshop, Vall de Nuria, Catalonia, Spain, 30 September–2 October*. Monografies techniques, Institut Cartogràfic de Catalunya, Barcelona, **8**.

HIRANO, A., WELCH, R. & LANG, H. 2003. Mapping from ASTER stereo image data: DEM validation and accuracy assessment. *ISPRS Journal of Photogrammetry & Remote Sensing*, **57**, 356–370.

HUTCHINSON, M. F. 1989. A new method for gridding elevation and streamline data with automatic removal of pits. *Journal of Hydrology*, **106**, 211–232.

KASSER, M. & EGELS, Y. 2001. Chapitre 3, Confection de modèles numériques de terrain et de surfaces. *In*: *Photogrammétrie numérique*. Lavoisier, Paris.

KILIAN, J., HAALA, N. & ENGLICH, M. 1996. Capture and evaluation of airborne laser scanner data. *In*: *International Archives of Photogrammetry and Remote Sensing, Vol. XXXI, Part B3, Vienna*. International Society for Photogrammetry and Remote Sensing (ISPRS), Beijing, 383–388.

OIMOEN, M. 2000. An effective filter for removal of production artifacts in U.S. Geological Survey 7.5-minute digital elevation models. *In*: *Fourteenth International Conference on Applied Geologic Remote Sensing*, Las Vegas, Nevada, 6–8 November.

OUTAL, S. 2006. *Quantification par analyse d'images de la granulométrie des roches fragmentées: amélioration de l'extraction morphologique des surfaces, amélioration de la reconstruction stéréologique*. PhD thesis, Ecole des Mines de Paris, France.

PCI 2003. *OrthoEngine V9.0 – User Manual*. PCI Geomatica, Ontario.

SOILLE, P. 1999. *Morphological Image Analysis. Principles and Applications*. Springer, Berlin.

SOILLE, P. 2004. Optimal removal of spurious pits in grid digital elevation models. *Water Resources Research*, **40**, W12509, doi: 10.1029/2004WR003060.

SOILLE, P., VOGT, J. & COLOMBO, R. 2003. Carving and adaptive drainage enforcement of grid digital elevation models. *Water Resources Research*, **39**, 1366, doi: 10.1029/2002WR001879.

STEVENS, N. F., GARBEIL, H. & MOUGINIS-MARK, P. 2004. NASA EOS Terra ASTER: volcanic topographic mapping and capability. *Remote Sensing of Environment*, **90**, 405–414.

TOUTIN, T. 2002. Three-dimensional topographic mapping with ASTER stereo data in rugged topography. *IEEE Transactions on Geoscience and Remote Sensing*, **40**, 2241–2247.

TOUTIN, T. 2004. Comparison of stereo-extracted DTM from different high-resolution sensors: SPOT-5, EROS-A, IKONOS-II & Quickbird. *IEEE Transactions on Geoscience and Remote Sensing*, **42**, 2121–2129.

TRINDER, J. C., WANG, P., LU, Y., WANG, Z. & CHENG, E. D. 2002. Aspects of digital elevation model determination. *In*: *Geospatial Theory, Processing and Applications. ISPRS Commission IV, Symposium, Ottawa, Canada, 9–12 July*. International Society for Photogrammetry and Remote Sensing (ISPRS), Beijing.

VINCENT, L. 1993. Morphological grayscale reconstruction in image analysis: applications and efficient algorithms. *IEEE Transactions on Geoscience and Remote Sensing*, **2**, 176–201.

WANG, P. 1998. Applying two dimensional Kalman filtering for digital terrain modelling. *In*: *The International Archives of Photogrammetry and Remote Sensing, Vol. XXXII, Part 4, ISPRS Commission IV – GIS Between Visions and Applications, Stuttgart, Germany,*

7–10 *September*. International Society for Photogrammetry and Remote Sensing (ISPRS), Beijing, 649–656.

WOODS, J. 1996. *The geomorphological characterisation of digital elevation models*. PhD thesis, Leicester University.

ZHANG, K., CHEN, S.-C., WHITMAN, D., SHYU, M.-L., YAN, J. & ZHANG, C. 2003. A progressive morphological filter for removing non-ground measurements from airborne LIDAR data. *IEEE Transactions on Geoscience and Remote Sensing*, **41**, 872–882.

The accuracy of ASTER digital elevation models: a comparison with NEXTMap Britain

M. HALL[1]* & D. G. TRAGHEIM[2]

[1]*Infoterra Ltd, Atlas House, 41 Wembley Road, Leicester LE3 1UT, UK*

[2]*British Geological Survey, Kingsley Dunham Centre, Keyworth, Nottingham NG12 5GG, UK*

*Corresponding author (e-mail: Michael.Hall@infoterra-global.com)

Abstract: In many overseas geological surveying projects an accurate elevation model is often required for analysis, image orthorectification, navigation and the generation of contours. Acquiring an accurate elevation model can be a difficult and expensive task. One possible solution is to generate a digital elevation model (DEM) from Advanced Spaceborne Thermal Emission and Reflection Radiometer (ASTER) satellite imagery. However, to fully understand the potential of ASTER DEMs the accuracy of these models needs to be established. The DEM was created using the Sulsoft ASTER DTM add-on ENVI module.

NEXTMap provides an ideal reference dataset for comparison. In this study the accuracy of an ASTER generated DEM was assessed for a 50 × 50 km area in central Wales. A total of 2.4 million points were compared.

Visual and statistical assessments were made, including profile and contour comparisons, allowing the spatial variation in accuracy to be explored. A mean vertical difference of −0.98 m and a standard deviation of *c.* 9 m were calculated. This suggests that 95% of the ASTER DEM points are within ±20 m of the NEXTMap DEM. Considering these accuracy levels, contours from ASTER can be generated at 40 m intervals.

The accuracy of ASTER DEMs

Digital elevation models (DEMs) have been successfully generated from stereo satellite imagery for some time, with the cross-track stereo SPOT1–4 being used from 1986 onwards. The launch of along-track stereo sensors allowed the images forming the stereo pairs to be collected within a short time separation and so overcoming temporal changes within the scene that are common in cross-track systems, such as variations in lighting, atmospheric conditions, cloud cover and vegetation. This high temporal correlation allows the automatic stereo-matching process to run more effectively and the resultant DEM to be generated more accurately. The Advanced Spaceborne Thermal Emission and Reflection Radiometer (ASTER) is only one of a series of platforms offering along-track stereo capability, including IKONOS and SPOT5, but the low cost of ASTER data coupled with its multispectral capability gives it a significant advantage over other sensors for geological applications.

Previous studies

In this study ASTER will be compared with NEXTMap DEMs for an area in Wales. There have been a number of accuracy studies assessing ASTER data (Kääb 2002; Hirano *et al.* 2003; Cuartero *et al.* 2004; Poli *et al.* 2004; San & Suzen 2005), but few have had access to such a high-quality DEM with as many comparison points. Accuracies are usually reported to be around 1 pixel (15 m) in root mean square error (RMSE) and standard deviation (SD), as shown in Table 1.

Commonly the RMSE and SD are the statistics used to measure DEM accuracy. In order for these figures to be interpreted by users of the DEM it is useful to give confidence levels (68/95/99%). However, it should be noted that assigning a confidence level to RMSE and SD based on reference to a DEM's true position is not possible unless the mean offset is accounted for or is zero or very close to it in relation to the size of the SD.

ASTER

ASTER is one of a number of sensors carried by the National Aeronautics and Space Administration's (NASA's) Terra spacecraft. It was launched in 1999 and offers wide spectral coverage, with a total of 14 bands in the visible and near infrared (VNIR), shortwave infrared (SWIR) and thermal infrared (TIR) sections of the electromagnetic spectrum, as shown in Table 2. Stereo images are acquired in band 3 using both the nadir (3N) and backward (3B) pointing telescopes at 15 m resolution, with the scene covering *c.* 60 × 60 km.

From: FLEMING, C., MARSH, S. H. & GILES, J. R. A. (eds) *Elevation Models for Geoscience.*
Geological Society, London, Special Publications, **345**, 43–53.
DOI: 10.1144/SP345.6   0305-8719/10/$15.00 © The Geological Society of London 2010.

Table 1. Summary table of ASTER DEM accuracy studies (after Hirano et al. 2003)

Study	Area and size	DEM post spacing (m)	No. of GCPs	No. of comparison points	RMSEz (m)	SD (m)	Minimum (m)	Maximum (m)	Mean (m)
Hirano et al. (2003)	Mt Fuji (24 × 21 km)	75	5 map points 1:25 000	51 map points 1:25 000	±26.3	–	–	–	–
Hirano et al. (2003)	Andes Mountains (55.5 × 57 km)	150	5 map points 1:50 000	53 map points 1:50 000	±15.8	–	–	–	
Hirano et al. (2003)	San Bernardino (22.5 × 22.5 km)	75	12 DGPS points	16 map points 1:24 000	±10.1	–	–	–	–
Hirano et al. (2003)	Huntsville (22.5 × 18 km)	30	8 DGPS points	239 776 posts (USGS DEM)	±14.7	–	–	–	
Poli et al. (2004)	Switzerland	–	46 map points 1:25 000	112 326 posts (25 m DEM)	±18.32	16.68	−84.49	67.89	−7.58
Cuartero et al. (2004)	Granada, southern Spain	30	15 map points	315 DGPS points	±12.6	12.5	–	–	−1.5
San & Suzen (2005)	Asarsuya River Basin, Turkey (20 × 10 km)	15	60 map points 1:25 000	DEM generated from 1:125 000 contours	–	17.69	−110.15	128.39	15.77

Table 2. *ASTER specifications*

Subsystem	Band no.	Spectral range (μm)	Spatial resolution (m)
VNIR	1	0.52–0.60	15
	2	0.63–0.69	
	3N	0.78–0.86	
	3B	0.78–0.86	
SWIR	4	1.600–1.700	30
	5	2.145–2.185	
	6	2.185–2.225	
	7	2.235–2.285	
	8	2.295–2.365	
	9	2.360–2.430	
TIR	10	8.125–8.475	90
	11	8.475–8.825	
	12	8.925–9.275	
	13	10.25–10.95	
	14	10.95–1.65	

NEXTMap Britain

NEXTMap Britain is a national high-resolution elevation dataset with a 5 m post spacing, generated from airborne infererometric synthetic aperture radar (InSAR) with a quoted accuracy of 1.0 m RMSE (Type 2 data). Two main elevation models make up the dataset, these are the digital surface model (DSM) and the digital terrain model (DTM). The DTM has undergone editing to remove cultural features and smaller areas of trees, whereas the DSM represents the unedited model. It is assumed in this study that the NEXTMap DEM is of a higher level of accuracy than the ASTER DEM and will be used as a reference dataset against which comparisons will be made.

DEM generation

ENVI was used to generate the DEM using the ASTER DEM extraction module. The scene chosen covers an area in Central Wales and is a level 1B processed image (Fig. 2). The topography on the area could be classed as low mountains. Nine control points were located across the scene (Fig. 1), extracted from OS 1:50 000 scale mapping. Although maps of greater accuracy are available for the test site, 1:50 000 scale maps were chosen to simulate the scale of maps that are

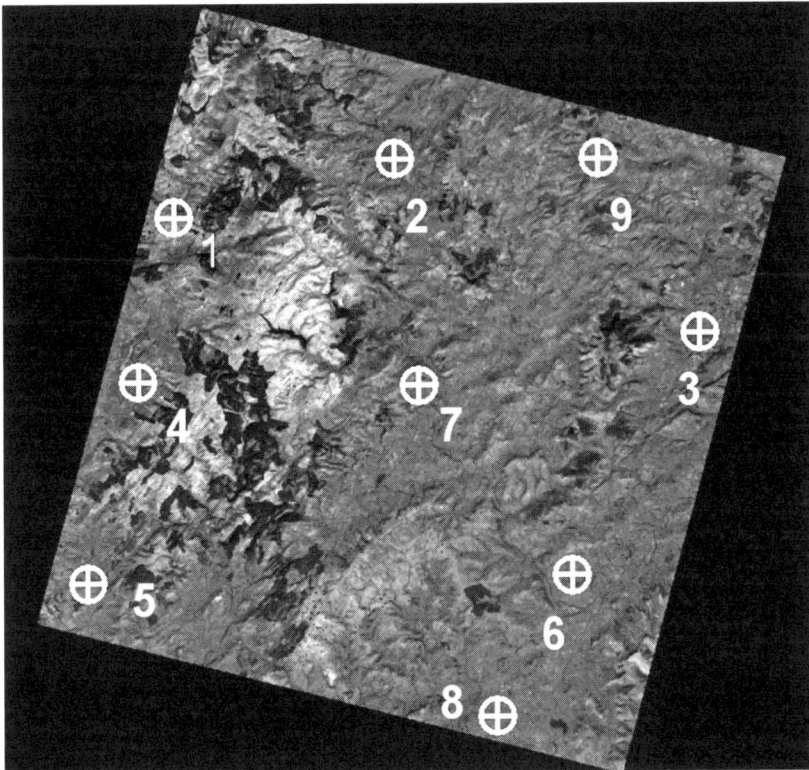

Fig. 1. Location of control points within the scene.

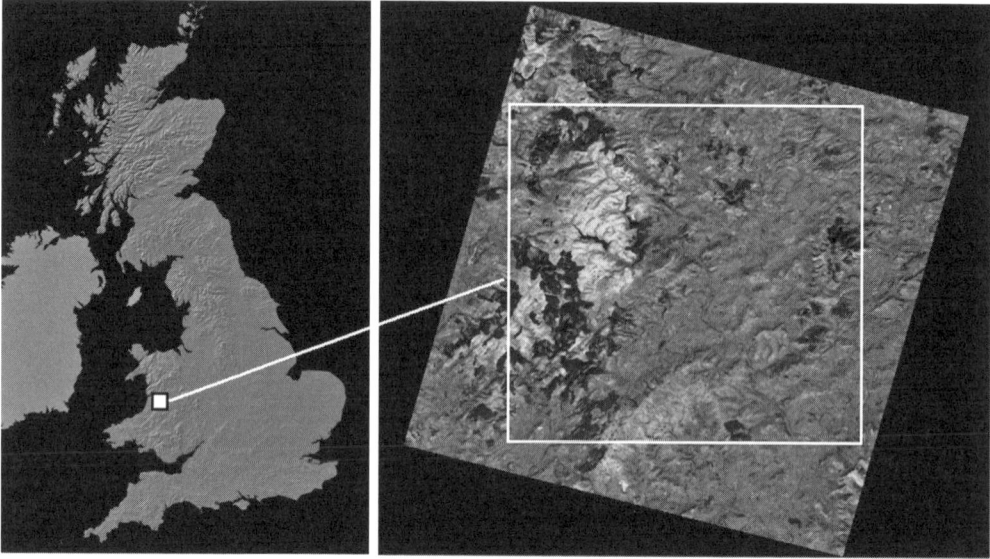

Fig. 2. ASTER image location.

typically available in overseas projects. Furthermore, it has been stated that '1:100 000 or 1:50 000 scale sources will probably provide sufficient accuracy' (Lang & Welch 1999). Ground control points (GCPs) were typically located on spot heights at road intersections with an even distribution across the imagery (Fig. 1), returning a maximum reported error of 3.3 pixels (Table 3). These values should be treated with caution as the manual states that they are not strictly accurate, so the emphasis was placed on locating the points correctly rather than trying to achieve a low RMSE value. The ENVI ASTER DEM extraction module expects co-ordinates to be given in UTM WGS84 datum, so the GCP co-ordinates were transformed from OSGB (Ordnance Survey of Great Britain) using an IMAGINE co-ordinate calculator before being entered.

The correlation score map produced within the ASTER DEM extraction module is shown in Figure 3. Darker colours indicate areas that have a low correlation between the two 3N and 3B images. These low correlation areas were found to correspond to forest, open water and moorland areas. The 30 m generated DEM (Fig. 4) was exported to IMAGINE and then reprojected with z values recalculated to OSGB.

The 5 m post spacing NEXTMap DEM was resampled to 30 m using the Degrade function within IMAGINE and then subtracted from the ASTER DEM to create a difference dataset. Statistics were calculated using SPLUS and visual comparisons made within ArcMap.

Table 3. *Summary report of GCP point accuracy*

GCP	Error X (pixels)	Error Y (pixels)	RMS (pixels)
1	−1.7100	0.1574	1.7172
2	1.9592	1.4835	2.4575
3	−2.1303	−1.9018	2.8557
4	3.2322	−0.3926	3.2560
5	−2.5019	−0.1770	2.5082
6	−1.3601	0.5820	1.4794
7	1.3620	0.3014	1.3949
8	1.7940	0.3254	1.8233
9	−0.6451	−0.3783	0.7478

Accuracy assessment

The main statistical comparison was carried out over a subset centred on the scene, as shown in Figure 5. Visually, there appear to be no apparent holes within the ASTER DEM. However, when the ASTER DEM is viewed as a shaded relief image and compared with the NEXTMap DEM, a terracing artefact effect can be seen on the steeper slopes (Fig. 6).

The results of the statistical comparison are shown in Table 4. The comparison between the NEXTMAP DSM and the ASTER DEM show that there is a mean difference of −0.98 m and a SD of

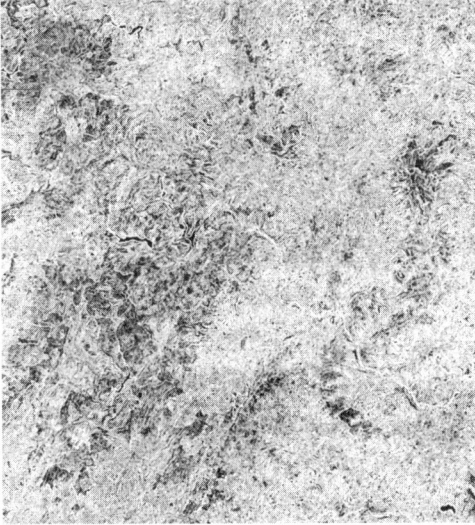

Fig. 3. Correlation score map; darker areas indicate lower correlation between the 3N and 3B image.

c. 9 m. Taking into account this mean offset, the results suggests that 68% of the ASTER DEM points are within 10 m of the NEXTMap DEM, 95% within 20 m and 99% within 30 m. Based on the 2.4 million comparison points, a RMSEz of ± 9 m was calculated, which compares more favourably than other accuracy assessment studies. The high minimum and maximum difference (-110 and 102 m) relate to outliers that exist within the dataset; the significance of these points is low, as shown by the histograms illustrating the spread of data in Figures 7 and 8. A negative shift in the mean can be observed but the magnitude of the differences follow the typical normal distribution.

The ASTER DEM has a slightly greater accuracy when compared with the NEXTMap DTM; this is likely to be related to the smoothing of surface features during the ASTER DEM generation process.

The spatial variation in the magnitude of elevation difference is shown in Figure 9. This has been calculated by taking the NEXTMap DSM elevation values from the ASTER and then colour coding them according to the magnitude of

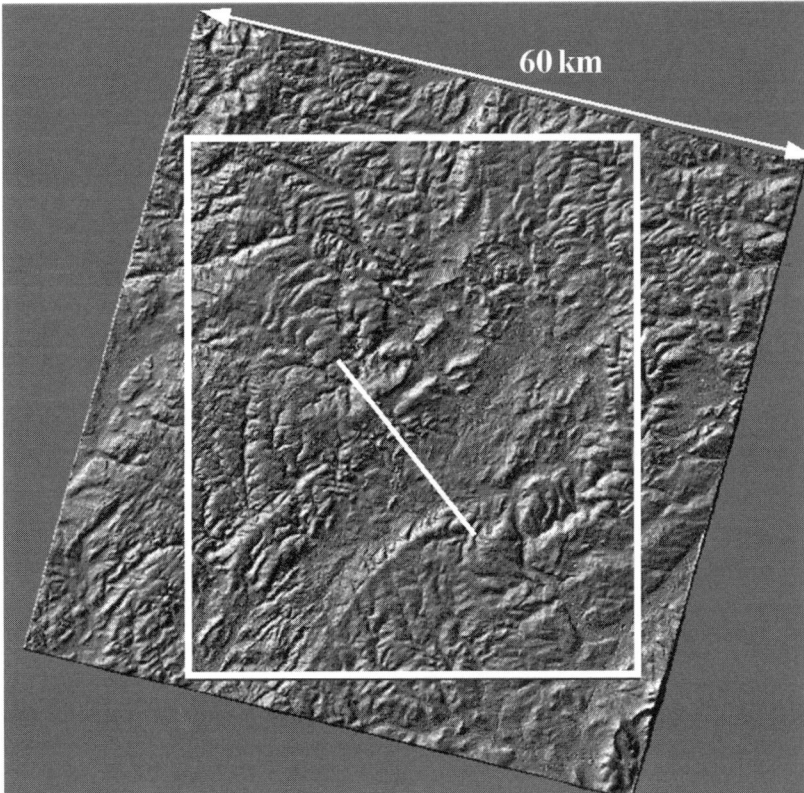

Fig. 4. Generated 30 m DEM.

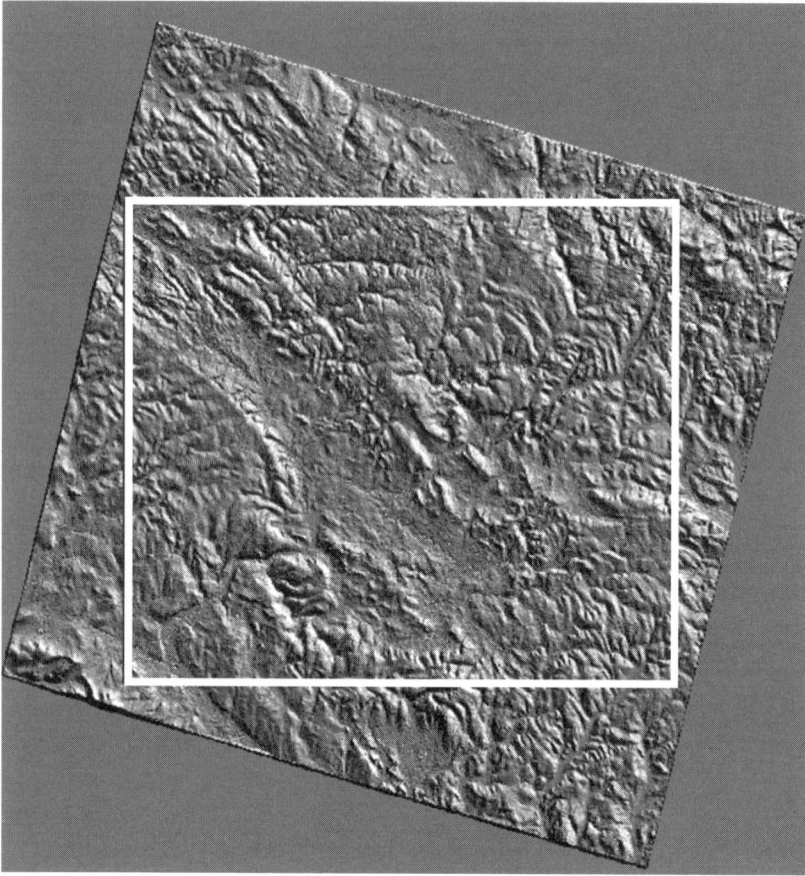

Fig. 5. ASTER generated 30 m DEM – shaded relief image; the main study area is shown by the box. Profile location is shown by the line.

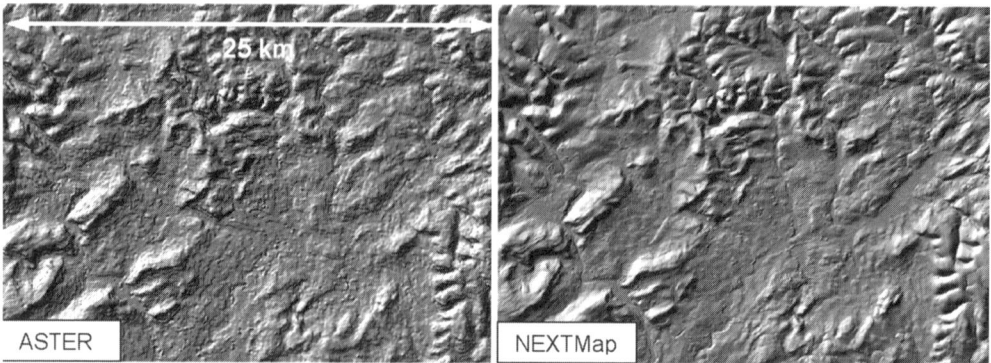

Fig. 6. Visual comparison between the ASTER and NEXTMap DEM. The ASTER DEM appears to show a terracing effect when viewed as a shaded relief image.

Table 4. *Accuracy assessment summary statistics*

	ASTER DSM*	ASTER DTM*	ASTER DSM subset area 1[†]	ASTER DSM subset area 2[‡]
Mean (m)	−0.98	−0.96	10.20	−6.38
Minimum (m)	−110.50	−111.64	−50.97	−71.42
Maximum (m)	102.17	94.63	73.01	51.74
SD (m)	9.012	8.73	5.77	8.41
RMSEz (m)	9.07		11.72	10.55

*2 441 463 points.
[†]101 136 points.
[‡]96 990 points.

elevation difference. Colder colours indicate areas where the ASTER DEM is lower than the NEXTMap DEM and warmer coloured areas those that are higher on the ASTER DEM. The most striking feature of the difference model is a cyclic banding pattern with alternating zones of positive and negative elevation differences. This spatial variation in accuracy does not correspond to variations in topography but is aligned parallel to the scan direction of the ASTER Image. It is difficult to be certain what is causing these areas of local offset but possible sources include initial pre-processing of the ASTER data, calibration problems or artefacts caused by the DEM extraction process.

To further investigate the DEM points within the areas of banding, two subsets were taken (shown in Fig. 9): one in the area where the ASTER DEM is higher than the NEXTMap DSM (subset area 1) and one where the ASTER DEM is lower (subset area 2). Table 4 shows the statistics for the two subset areas. In subset area 1 the mean difference is 10.2 m and in area 2 the mean difference is −6.4 m, this shift is also illustrated by the histogram of pixel differences (Figs 10 & 11). Area 1 has a

RMSEz of ± 11.7 m and area of 2 ± 10.6 m. Although these areas represent the lowest accuracy zones in comparison to NEXTMap, the statistical accuracy is comparable to other studies (namely Poli *et al.* 2004).

Looking more closely at the difference image (Fig. 12), we can start to find the causes of local variations between the ASTER and NEXTMap datasets. Valley bases are higher on the ASTER DEM and areas of trees generally lower, suggesting a general smoothing of features. However, there are differences that cannot be easily explained by topographical variations in terrain characteristics and may be simply a result of the ASTER DEM extraction process.

There is also a tendency for the ASTER DEM to be higher than the NEXTMap DSM on NW- and north-facing valley sides, which may be due to a shadowing effect, influencing the ASTER DEM extraction routine.

A profile comparison across a 2 km-transect (Fig. 3) generally shows a good correspondence between the ASTER DEM and the NEXTMap DSM (Fig. 13). The terracing effect discussed

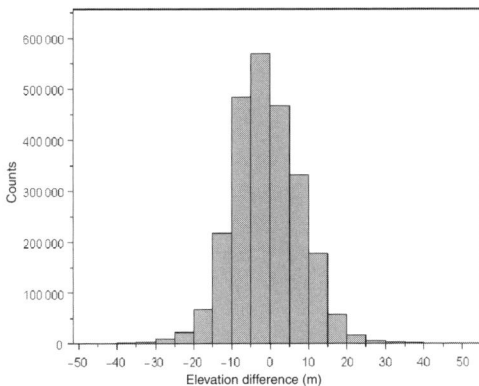

Fig. 7. Histogram showing the elevation differences between the ASTER and NEXTMap DSM, as pixel counts. Date grouped into 5 m classes. Approximately 2.4 million points.

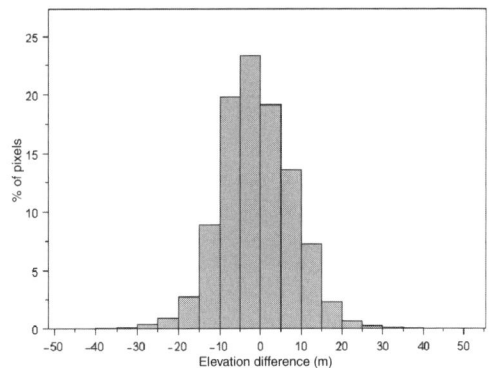

Fig. 8. Histogram showing the elevation differences between the ASTER and NEXTMap DSM, as percentages. Data grouped into 5 m classes. Approximately 2.4 million points.

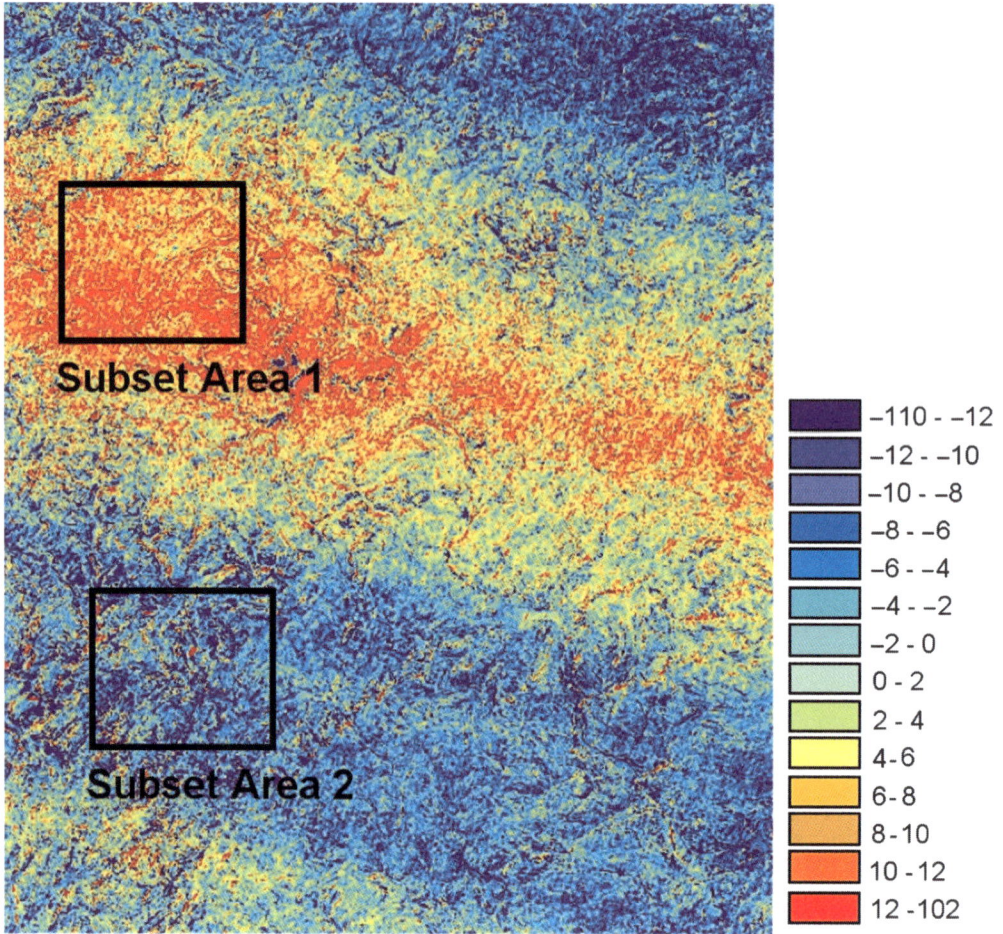

Fig. 9. Difference (in m) between the ASTER and NEXTMap DSM; colder colours indicate areas where the ASTER DEM is lower than the NEXTMap DEM.

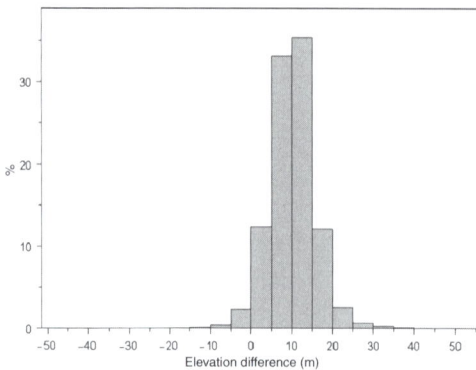

Fig. 10. Histogram showing the elevation differences between the ASTER and NEXTMap DSM as a percentage of pixels for subset area 1; data grouped into 5 m classes.

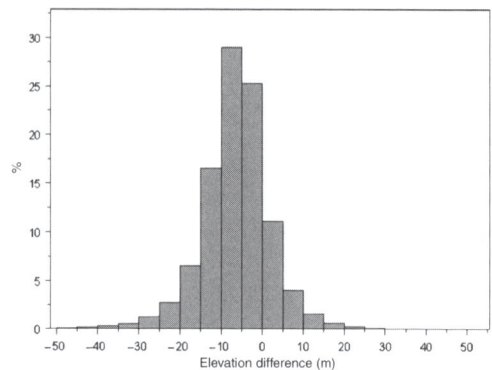

Fig. 11. Histogram showing the elevation differences between the ASTER and NEXTMap DSM as a percentage of pixels for subset area 2; data grouped into 5 m classes.

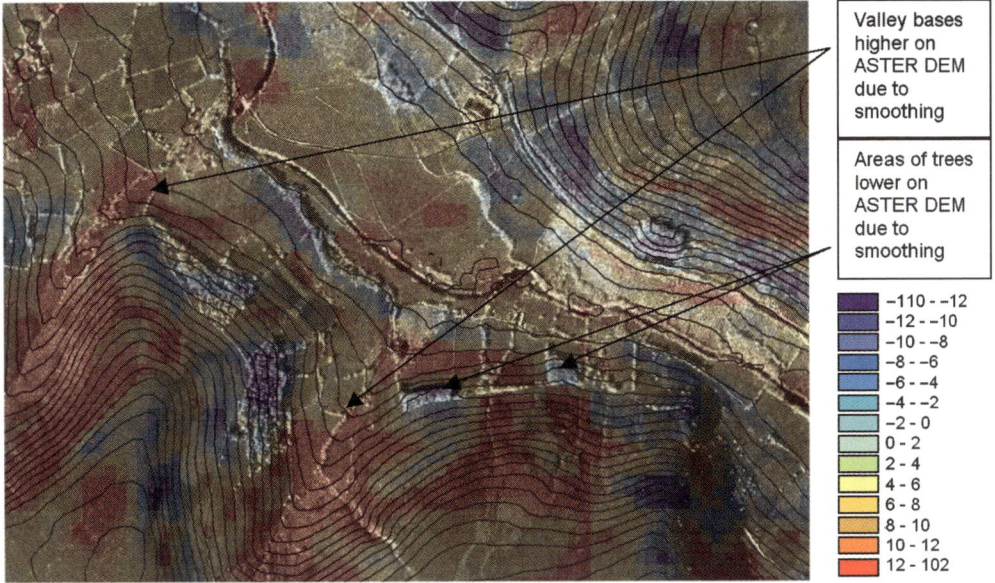

Fig. 12. Difference image illustrating the spatial pattern of differences (in m) between the ASTER and NEXTMap DSM; NEXTMap contours overlain to illustrate relief variations.

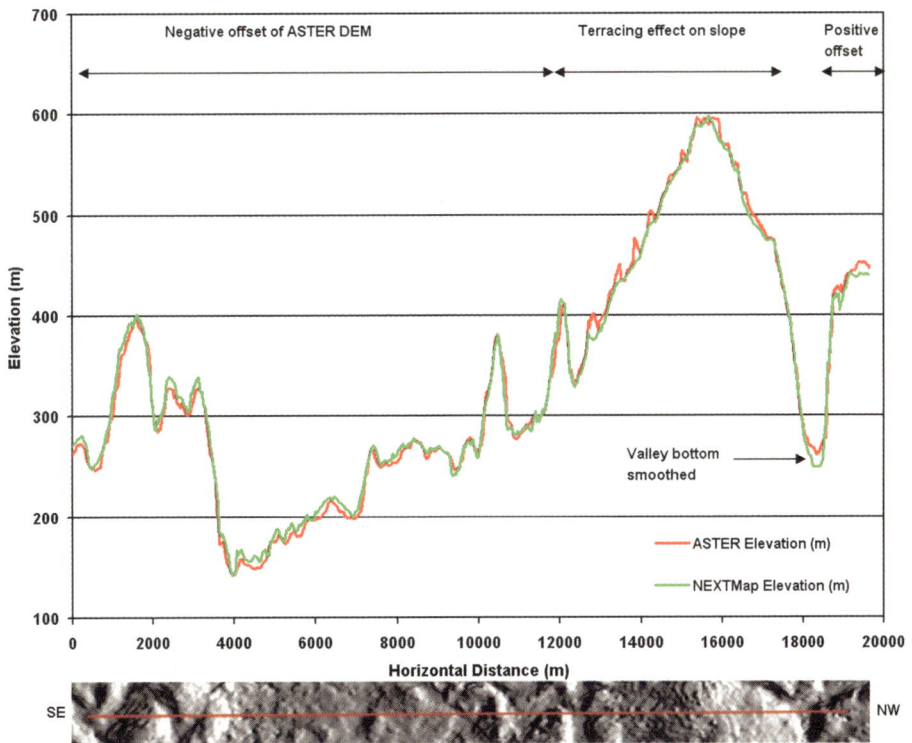

Fig. 13. Profile across the ASTER and NEXTMap DSM; shaded relief image is shown to illustrate terrain characteristics.

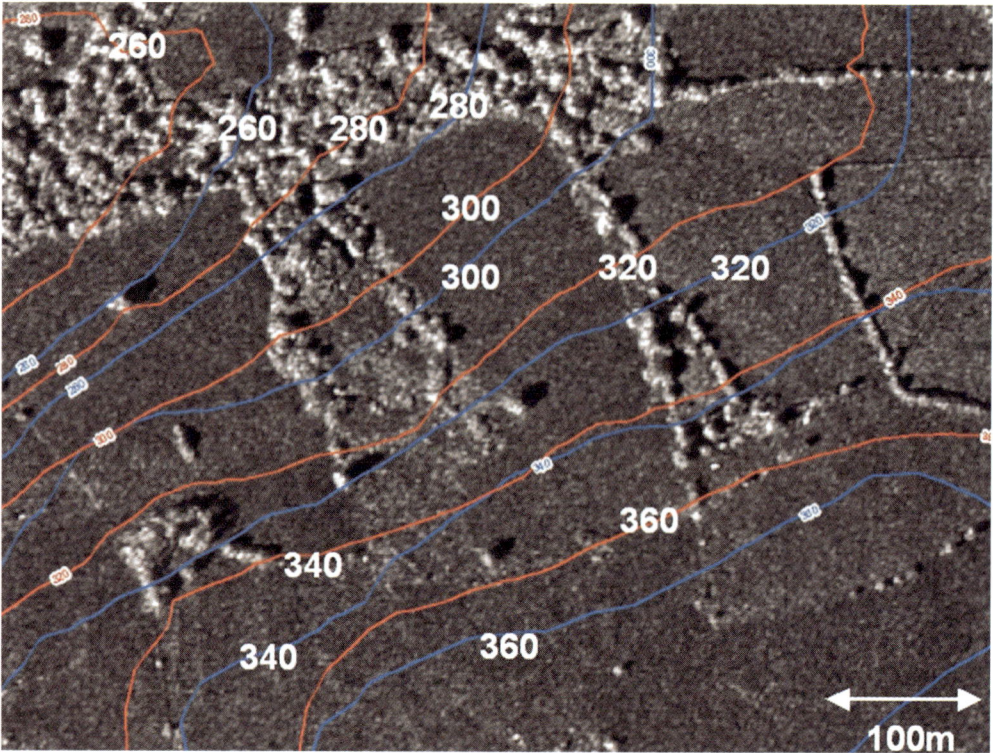

Fig. 14. A comparison of generated 20 m contours from the ASTER DEM and NEXTMap DSM for a regular sloping hillside, overlain onto the NEXTMap orthorectified radar image. NEXTMap contours in blue, ASTER contours in red.

earlier can be seen on one of the slopes, together with a smoothing of the valley bottoms. For the first 10 km of the profile the ASTER DEM can be seen to follow closely to the NEXTMap DEM but with a negative offset.

Contour comparison

Contours, automatically generated for the ASTER DEM and NEXTMap DSM, are shown in Figure 14. Although not superimposing perfectly, they show a reasonable correspondence in location. Valleys tend to be less well defined on the ASTER DEM and a degree of smoothing was noted on spurs. It was found that viewing 20 m ASTER contours at 1:100 000 or even 1:50 000 scale is possible without showing significant disparity with the NEXTMap contours. Assuming a 4× SD rule to define contour intervals, a 40 m-contour interval can be deemed appropriate.

Conclusions

It has been demonstrated that ASTER DEMs can be generated with a level of accuracy suitable for

fulfilling many geological and mapping application in international projects, provided that adequate ground control is available. In comparison to the NEXTMap DSM, the calculated accuracy was found to be comparable to or better than previous assessments of ASTER DEMs.

The marked spatial variation in accuracy, shown by areas of positive and negative offset aligned parallel to the scan direction, represents the lowest accuracy zones in comparison to NEXTMap. However, the statistical accuracy of these areas is still comparable to other studies, namely Poli *et al.* (2004), and clearly warrants further investigation.

References

CUARTERO, A., FELICISIMO, A. M. & ARIZA, F. J. 2004. Accuracy of DEM generation from Terra–Aster stereo data. *IAPRS&SIS*, **35**, 225–260.

HIRANO, A., WELCH, R. & LANG, B. 2003. Mapping from ASTER stereo image data: DEM validation and accuracy assessment. *ISPRS Journal of Photogrammetry and Remote Sensing*, **57**, 356–370.

KÄÄB, A. 2002. Monitoring high-mountain terrain deformation from repeated air- and spaceborne optical data: examples using digital aerial imagery and

ASTER data. *ISPRS Journal of Photogrammetry and Remote Sensing*, **57**, 39–52.

LANG, H. & WELCH, R. 1999. *Algorithm Theoretical Basis Document for ASTER Digital Elevation Models, Version 3.0*. Jet Propulsion Laboratory, Pasadena, CA, 69.

POLI, D., REMONDINO, F. & DOLCI, C. 2004. Use of satellite imagery for DEM extraction, landscape modelling and GIS applications. Available online at: http://www.photogrammetry.ethz.ch/general/persons/fabio/thai_DTM.pdf.

SAN, B. T. & SUZEN, M. L. 2005. Digital elevation (DEM) generation and accuracy assessment from ASTER stereo data. *International Journal of Remote Sensing*, **26**, 5013–5027.

The use of NEXTMap Britain for geological surveying in the Vale of York

M. HALL[1]*, A. H. COOPER[2], J. FORD[2], S. PRICE[2] & H. BURKE[2]

[1]*Infoterra Ltd., Atlas House, 41 Wembley Road, Leicester LE3 1UT, UK*

[2]*British Geological Survey, Kingsley Dunham Centre, Keyworth, Nottingham NG12 5GG, UK*

*Corresponding author (e-mail: Michael.Hall@infoterra-global.com)

Abstract: The NEXTMap Britain digital elevation model (DEM) has opened many new opportunities that considerably help and enhance the way we undertake our geological mapping of bedrock, structure, and superficial and artificial deposits. The dataset has been successfully integrated into the digital and conventional mapping workflows of the Vale of York mapping team.

A variety of visualization and analysis techniques have been applied throughout the mapping process. These techniques include an initial appraisal of NEXTMap with a comparison to existing geological mapping to define the field mapping strategy and site-specific manipulation using Tablet PCs.

NEXTMap interpretation has made an important contribution to the understanding of the extensive glacial and proglacial deposits found in the Vale of York; such as sand bodies resting on lake deposits, and identifying details within morainic and alluvial complexes. For bedrock mapping, NEXTMap has been used to identify landform features that relate to the underlying geology, such as breaks in slope, the extent of escarpments, hillcrests and dip slopes, to provide an overview of the landscape and to save time in mapping out features in the field. Techniques have also been developed to automatically generate these landform features.

The dataset has also been used to identify areas where landsliding has occurred, for the accurate mapping of artificial ground and as a key surface for three-dimensional (3D) geological modelling.

The York district (Sheet 63), originally surveyed in the 1880s, is currently being resurveyed as part of a national targeted revision programme; the Selby district (Sheet 71) has just been completed. Modern geological mapping has involved bringing together a range of digital datasets and the move towards a digital mapping workflow. The multidisciplinary work is integrating existing British Geological Survey map data holdings, geochemical information (G-Base), borehole records, seismic data, digital elevation models (DEMs), aerial photography interpretations and three-dimensional (3D) modelling techniques with traditional field-based geological surveying. The NEXTMap elevation model is a key dataset used extensively throughout the mapping workflow.

This paper focuses on the contribution that NEXTMap has made to the digital mapping workflow, including the processing techniques used for enhancing the dataset for geological feature collection. It also briefly covers its use within the digital field data capture system and for the production of 3D geological models.

Study area

The Vale of York is a broad area of low-lying and generally gently undulating land in the NE of England constrained between the Yorkshire Wolds and Howardian Hills to the east and the Pennines to the west (Fig. 1). The bedrock geology in the Vale consists of Triassic sandstones and mudstones bounded by Jurassic and Cretaceous rocks to the east and with Permian and Carboniferous rocks to the west (Fig. 2a). Overlying the bedrock geology there is a varied cover of superficial deposits consisting of glacial till, outwash sands and gravels, glaciolacustrine clay, peat and alluvium (Fig. 2b). Their distribution is strongly influenced by the advance and retreat of ice during the Devensian glaciation. Detailed descriptions of the geology of the Vale of York and surrounding areas can be found in Fox-Strangways (1884), Gaunt (1981, 1994), Powell *et al.* (1992), Cooper & Burgess (1993) and Cooper & Gibson (2003).

Terrain modelling for geological applications

In temperate environments such as the UK, extensive vegetation cover and soil development results in little exposure of the underlying geology. Geological mapping relies on using the interpretation of landform features to tie together information obtained from exposures, field brash and auger holes. Traditionally, these features are

From: FLEMING, C., MARSH, S. H. & GILES, J. R. A. (eds) *Elevation Models for Geoscience.*
Geological Society, London, Special Publications, **345**, 55–66.
DOI: 10.1144/SP345.7   0305-8719/10/$15.00 © The Geological Society of London 2010.

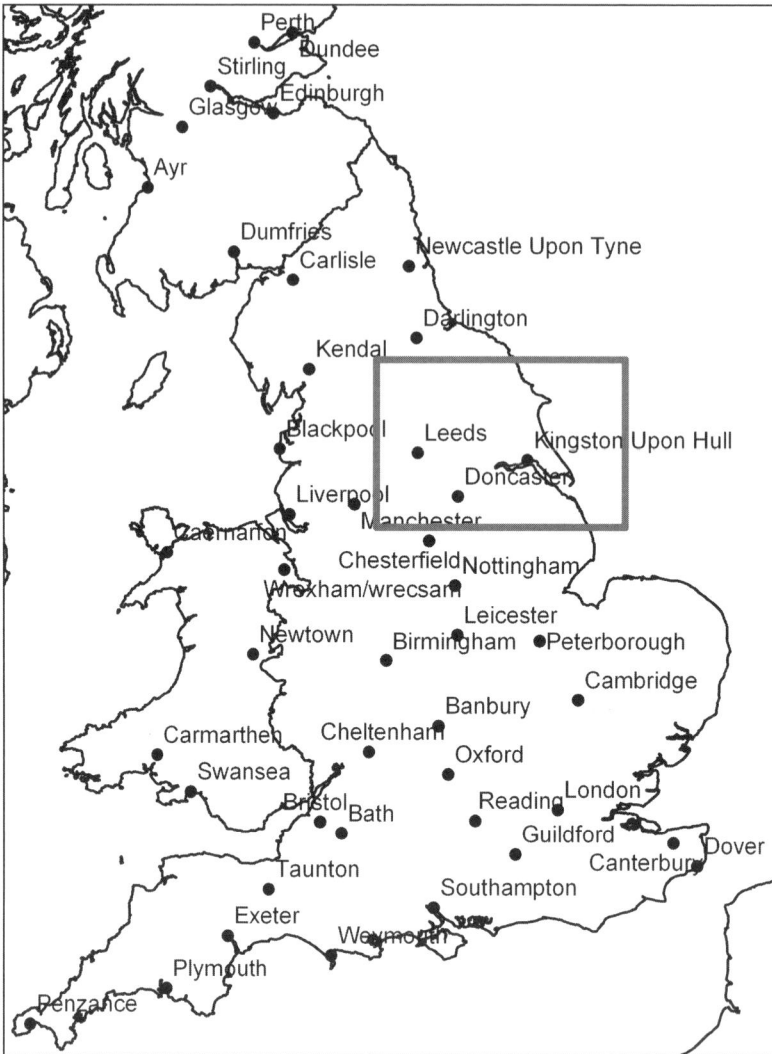

Fig. 1. Location map of the Vale of York and surrounding areas shown in Figure 2a & b.

identified in the field or from hard-copy aerial photographs using a stereoscope. However, both of these methods can be time consuming, and the interpretation of aerial photographs can be difficult owing to inherent distortions and scale variations present within the imagery. Modern methods such as digital stereo aerial photograph interpretation and the use of a digital elevation model (DEM) are the two main remote sensing techniques that can enhance the efficiency and accuracy of the landform feature capture process.

Elevation models have been used for some time as a useful tool for regional and local geological applications (Onorati *et al.* 1992; Pike 1992; Bonham-Carter 1994) and landform mapping

(Smith & Clark 2005). Focused studies have made use of aerial photography or LiDAR (Light Detection And Ranging) derived models, but at a relatively high cost, with contour-derived models being applied for regional applications. With the availability of NEXTMap, elevation models have become routinely used for geological applications within the British Geological Survey (Hall *et al.* 2004; Bradwell *et al.* 2007) and more widely (Smith *et al.* 2006). The combination of high-resolution and national coverage is ideally suited to geological mapping and modelling applications.

The NEXTMap dataset for England and Wales was collected by InterMap Technologies between 2002 and 2003 using Interferometric Synthetic

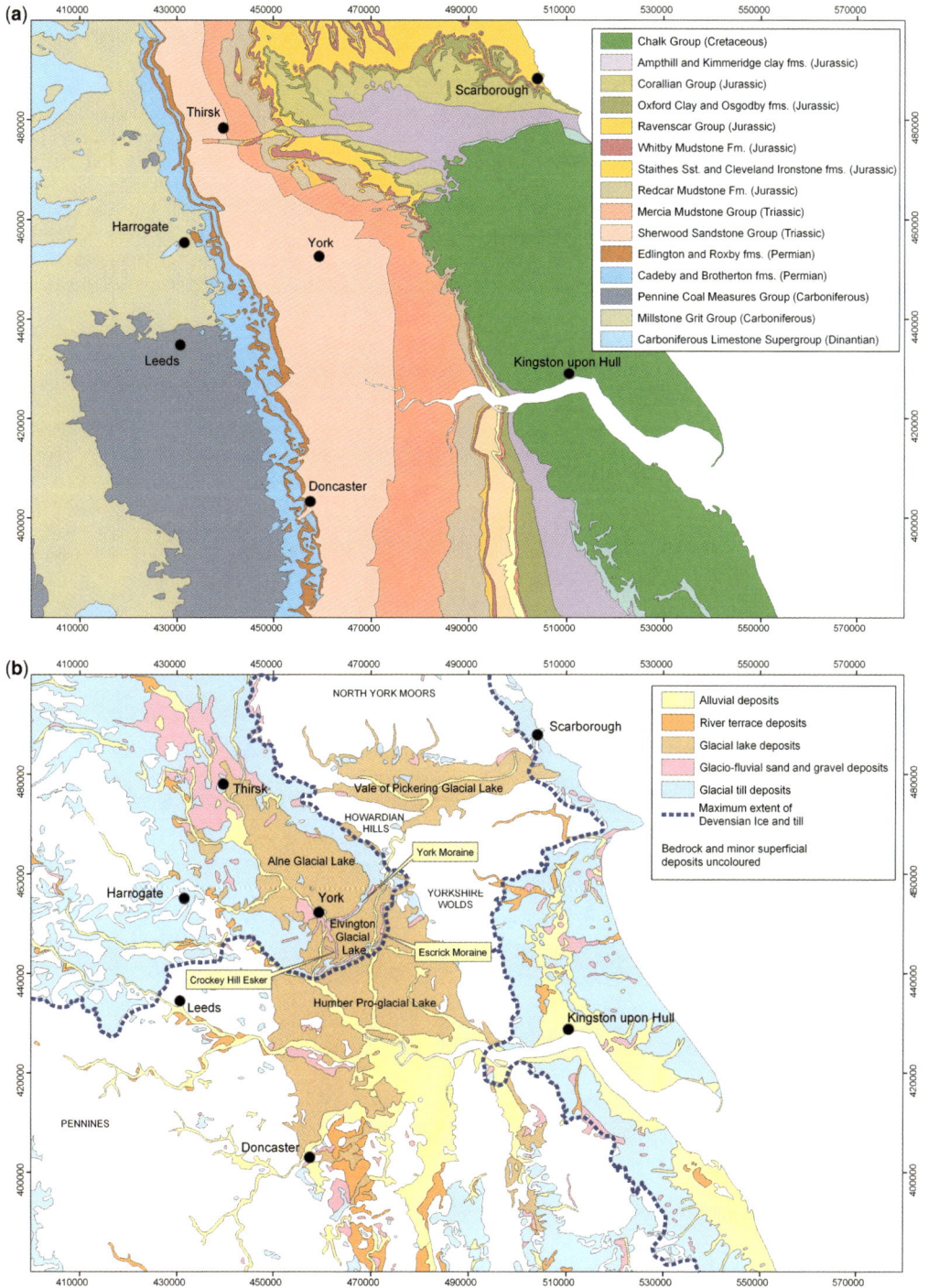

Fig. 2. (**a**) Simplified bedrock geology of the Vale of York and surrounding area extracted from the British Geological Survey 1:625 000 scale digital bedrock geology dataset. The Jurassic stratigraphy shown in the key is that for Yorkshire and, for clarity, a few of the thinner Jurassic and Triassic units are unnamed or not shown. (**b**) Simplified diagram of Devensian and Pre-Devensian deposits in the Vale of York and surrounding areas extracted from the British Geological Survey 1:625 000 scale digital superficial geology dataset. Devensian limit in the Vale of York based on borehole data and recent surveying.

Aperture Radar (InSAR), producing a 5 m post spacing DEM with a quoted vertical accuracy of 1.0 m root mean square error (RMSE) for Type 2 data (Intermap Technologies 2007). Two main elevation models make up the dataset, these are the digital surface model (DSM) and the digital terrain model (DTM). The DSM represents the unedited model. The DTM has undergone editing to remove cultural features and smaller areas of trees, embankments have been enhanced and the model has been modified to make it hydrologically correct for drainage.

Quaternary evolution of the Vale of York

The geological evolution of the Vale of York during the Quaternary produced a distinctive range of sediments and landforms that characterize the present-day area (Cooper & Burgess 1993; Gaunt 1994). The superficial deposits present in the Vale of York are largely a result of the last Devensian glaciation between about 115 000 and 10 000 years ago, when ice covered most of northern Britain (Cooper & Gibson 2003). The advance of

Vale of York ice and its subsequent retreat created an assemblage of ice-contact, ice-marginal and proglacial deposits. The main morphological features can be seen in the NEXTMap DTM (Fig. 3).

The Pennine valleys were glaciated as far south as Leeds and a tongue of ice occupied the Vale of York, advancing as far south as the Escrick Moraine (Fig. 4a). At the same time, the North Sea ice advanced to Norfolk, blocking the drainage out through the Humber Gap. In front of the ice, fluvioglacial outwash deposits and proglacial lake deposits were formed in the dammed preglacial valley system.

As the ice advanced to the Devensian maximum, forming the Escrick Moraine, it overrode many of the proglacial deposits. The ice front then retreated progressively northwards with intervening stillstands depositing the Crockey Hill Esker (Fig. 4a), the York Moraine (Fig. 4b) and the Flaxby–Tollerton Moraine (Fig. 4c) (Cooper & Burgess 1993). The York and Flaxby–Tollerton moraines represent stillstands in the ice margin where the supply of sediment-laden ice was in equilibrium with the degree of melting or wasting.

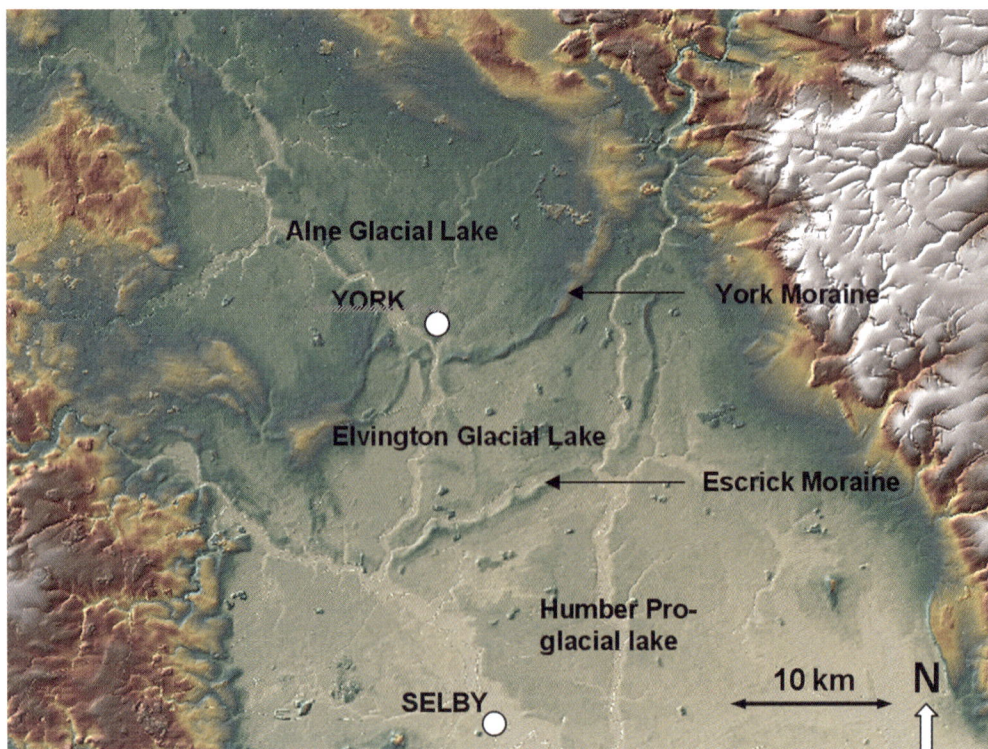

Fig. 3. The main morphological features of the Vale of York NEXTMap DTM, shaded and colour coded according to elevation. Light green colours represent low elevation, rising through yellow and brown to white, representing high elevation. Elevation data from Intermap Technologies.

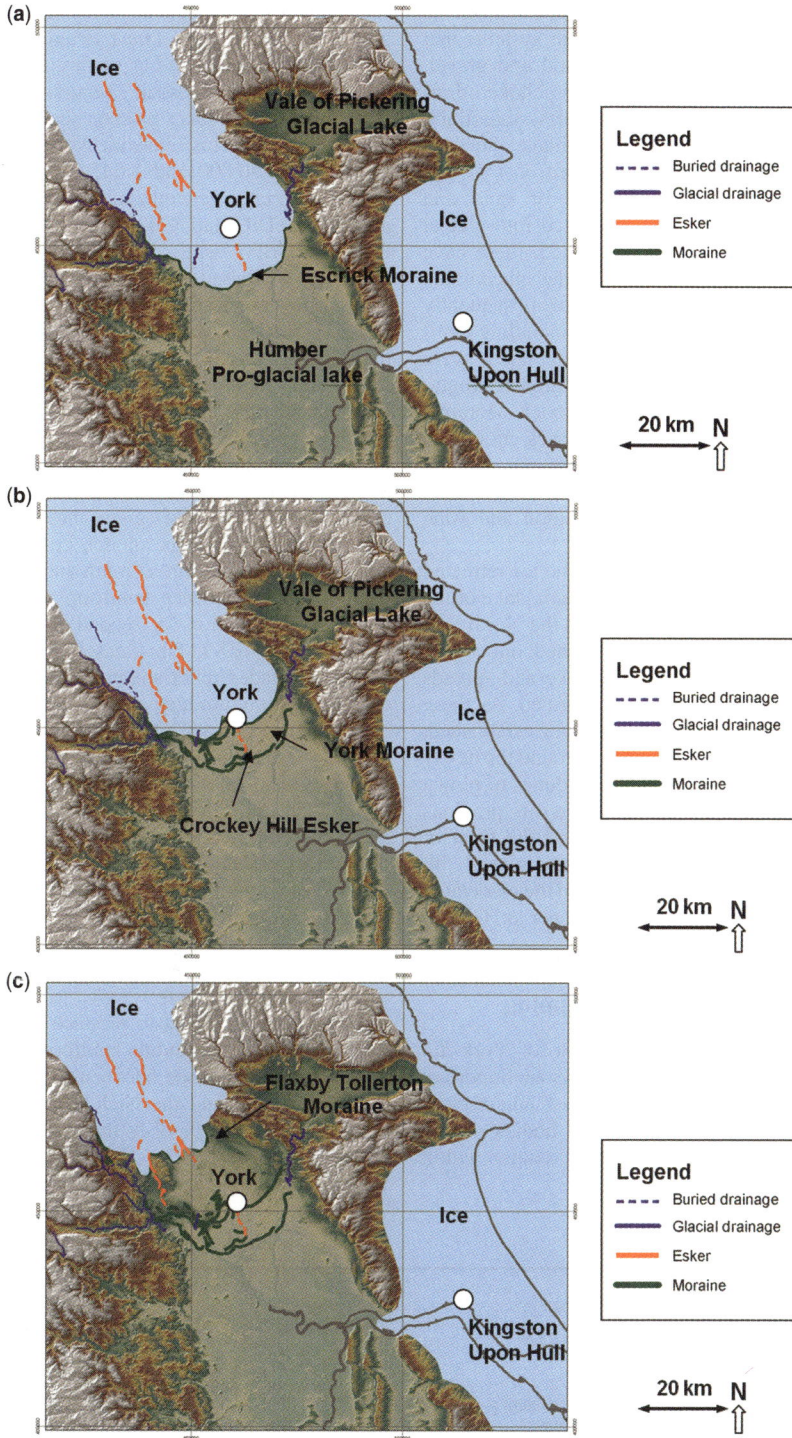

Fig. 4. Glacial landforms in the Vale of York and their relationships to stages of ice advance and retreat during the Devensian. (**a**) Ice position during Devensian Maximum, forming the Escrick Moraine. (**b**) Ice retreat and formation of the York Moraine during stillstand. (**c**) Ice retreat and formation of the Flaxby–Tollerton Moraine. Elevation data from Intermap Technologies.

The vast amounts of meltwater draining from the ice sheet formed subglacial drainage systems that commonly became choked with sand and gravel. Upon melting of this ice sheet, these choked drainage systems were left as ridges (eskers) of partially disturbed deposits. Where the drainage emerged from the ice sheet (at the sides or in front) it commonly deposited terraces or fans of sand and gravel. Where the drainage disgorged into glacial lakes, fans of sand and gravel formed; these have an upper surface approximating to the glacial lake water level. Some of these fans were subsequently buried by glaciolacustrine deposits, including laminated clay. Three separate glacial lakes were formed, the largest proglacial lake (Humber proglacial lake) being in front of the Escrick Moraine, another between the York and Escrick moraines, and another to the north of the York moraine; the deposits formed in these lakes are, from south to north, the Hemingbrough, Elvington and Alne glaciolacustrine formations (Fig. 5).

After the Devensian glaciation, the ice retreated from the Humber Gap and the proglacial lake of the Vale of York drained eastwards into the North Sea. Extensive sand deposits were washed out across the floor of the recently drained lake and spreads of sand with a little gravel were formed. As these deposits dried and the drainage became established, exposed sand deposits were blown around the newly emerged lake bed forming subdued dunes of blown sand. Much of the drainage followed its previous course into the Vale, such as around the front of the Escrick Moraine, cutting into the glacial till, the glaciofluvial outwash terraces and the associated glacial lake deposits (Fig. 5) (Cooper & Gibson 2003).

NEXTMap and superficial mapping

The glacial deposits within the vale of York are extensive and NEXTMap has made an important contribution to the understanding of lake-bottom elevations, the recognition of sand bodies resting on lake deposits, and identifying details within morainic and alluvial complexes.

The British Geological Survey (BGS) produces geological linework of Great Britain at a number of scales in digital and hardcopy format and in Superficial and Bedrock versions. Geological mapping typically takes place at scales of 1:10 000 or 1:20 000, and this linework is then produced in 1:50 000, 1:250 000 and 1:625 000 versions with the appropriate generalizations. These datasets are named DiGMap GB-50, 250 and 625, with DiGMap GB50 being the standard scale of the national sheet-based printed maps.

To allow comparison with the current DiGMap GB-50 scale geological linework, the NEXTMap data are processed to enhance the topographical information held within the dataset. A number of steps are involved with this enhancement process, which is undertaken in the ArcMap environment. First, the terrain model is artificially lit using the Spatial Analyst extension. This artificial lighting effect, or shaded relief, gives the appearance of sunlight illuminating the model, creating shadows and highlights. The sun azimuth and elevation can be modified to enhance landform features with a particular orientation. A second semi-transparent copy of the NEXTMap dataset is superimposed above the shaded relief and colour ramped according to elevation, producing a combined shaded and colour ramped image. The pre-revision linework can then be checked against the landforms displayed in the hill-shaded and colour ramped image. The features are typically subtle and individual colour ramp intervals of 10–15 cm were found to be most effective in identifying features. Fieldwork was then targeted in areas where there was found to be a discrepancy between the landform features and the existing linework. It was established that the DSM model was more useful than then DTM as the algorithm used to remove trees and culture features from the DSM to produce the DTM was found to also remove subtle landform features.

The correspondence between the existing geological linework and the NEXTMap DSM model is shown in Figure 6a & b; numerous features are clearly identifiable. The Crockey Hill Esker is visible as an elongated ridge aligned north–south;

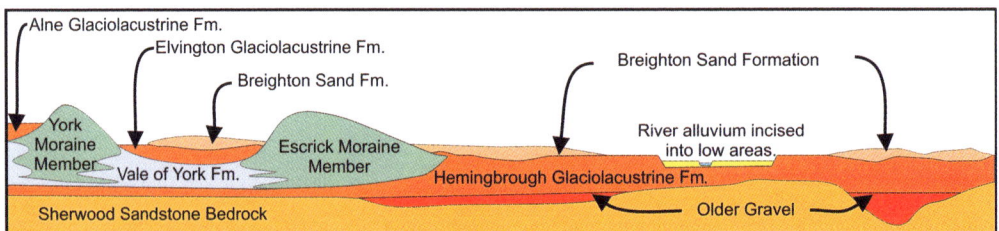

Fig. 5. Cross-section through superficial deposits in the Vale of York, illustrating the relationship between the units.

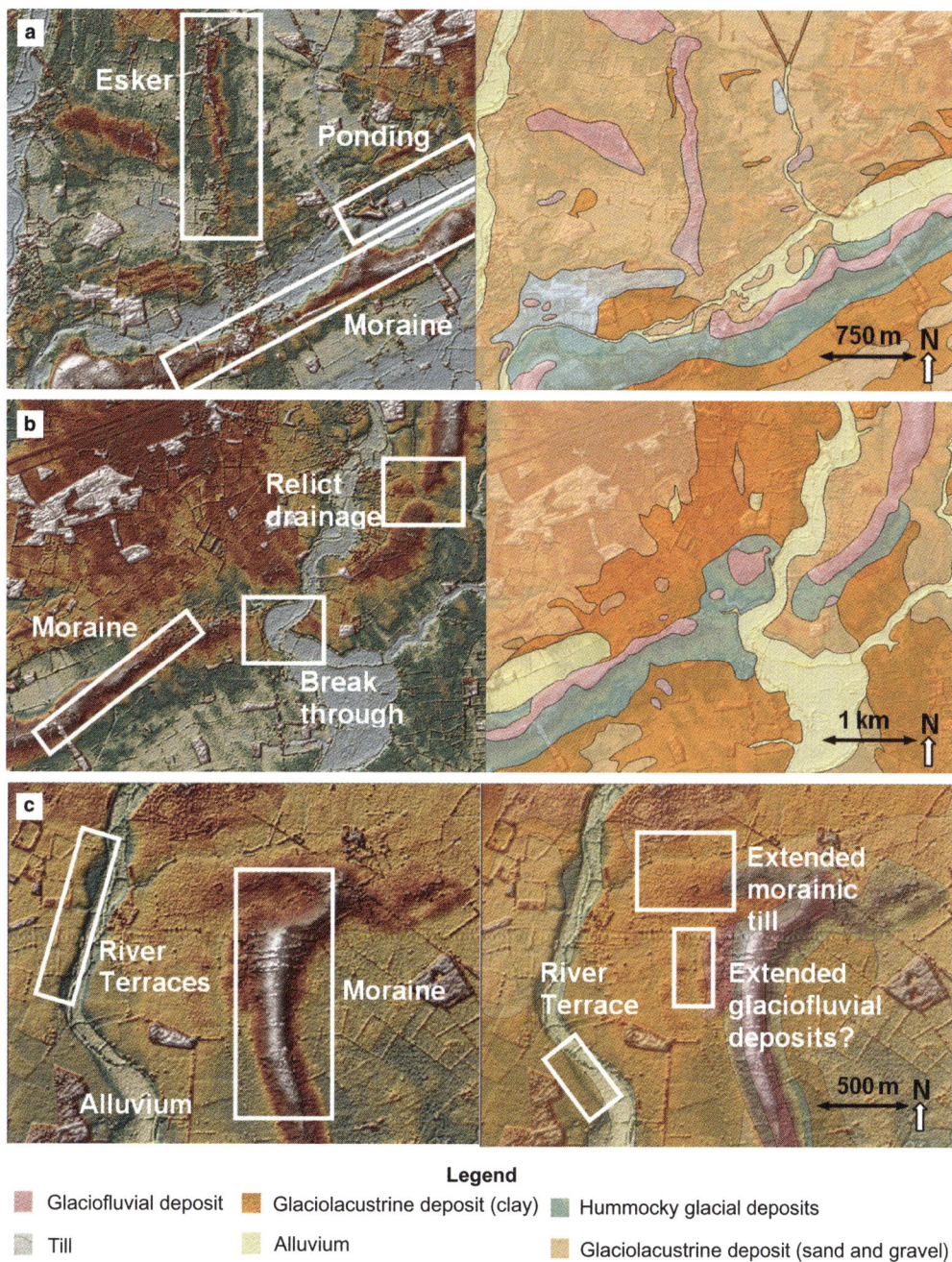

Legend

Glaciofluvial deposit	Glaciolacustrine deposit (clay)	Hummocky glacial deposits
Till	Alluvium	Glaciolacustrine deposit (sand and gravel)

Fig. 6. (a) Crockey Hill Esker, NEXTMap shaded and colour ramped DSM. Blue 5 m, yellow 10 m, white 30 m elevation. Existing DiGMap 1:50 000 scale superficial geology linework shown in the right-hand panel in semi-transparent format with underlying colour ramped and shaded NEXTMap DSM. Elevation data from Intermap Technologies. (**b**) Escrick Moraine centre. NEXTMap shaded and colour ramped DSM. Blue 5 m, yellow 10 m, white 30 m elevation. Existing DiGMap 1:50 000 scale superficial geology linework shown in the right-hand panel in semi-transparent format with underlying colour ramped and shaded NEXTMap DSM. Elevation data from Intermap Technologies. (**c**) Escrick Moraine north. NEXTMap shaded and colour ramped DSM Blue 5 m, yellow 10 m, white 30 m elevation. Existing DiGMap 1:50 000 scale superficial geology linework shown in the right-hand panel in semi-transparent format with underlying colour ramped and shaded NEXTMap DSM. Elevation data from Intermap Technologies.

sand and gravel deposits form several raised areas; and the Escrick Moraine in the southern part of the image is striking (Fig. 6a). Localized ponding caused by the moraine restricting draining in this part of the Vale has resulted in an area of alluvium in the low-lying areas behind the ridge. Figure 6b shows the area of high ground formed by the central part of the Escrick Moraine. A relict drainage channel can be identified, which is thought to represent drainage through the moraine prior to the break through by the River Derwent. The Derwent, with its associated floodplain deposits, can be seen in the centre of the image. Some areas show a clear discrepancy between the landform features present on the NEXTMap DSM and the existing mapping for the northern extent of the Escrick Moraine. In Figure 6c, a possible river terrace, not present on existing mapping, can be identified on the western bank of the Derwent; there are also areas where the morainic till and the glaciofluvial deposits could be extended.

The NEXTMap dataset has considerable advantages over contour-derived models in the low relief areas of the Vale of York. The wide spacing of contours was found to result in interpolation artefacts and the smoothing of subtle features. As the NEXTMap model is based on a radar signal collected at regular intervals along the ground surface, the model is particularly useful for the interpretation of superficial deposits in river floodplains and other low relief areas.

NEXTMap and bedrock mapping

An extensive outcrop of Jurassic and Cretaceous strata forms the eastern limit to the Vale of York and the prominent topographical feature of the Yorkshire Wolds. This area of higher ground is relatively well featured compared to the lower ground with its superficial deposit cover. The NEXTMap dataset was used for a number of aspects of bedrock mapping. Prior to fieldwork, a study of the DSM allowed breaks in slope to be recognized and interpreted in the context of the local geological succession. Discrepancies between existing geological mapping and the NEXTMap interpretation were targeted and prioritized for subsequent fieldwork, resulting in a more effective use of field resources.

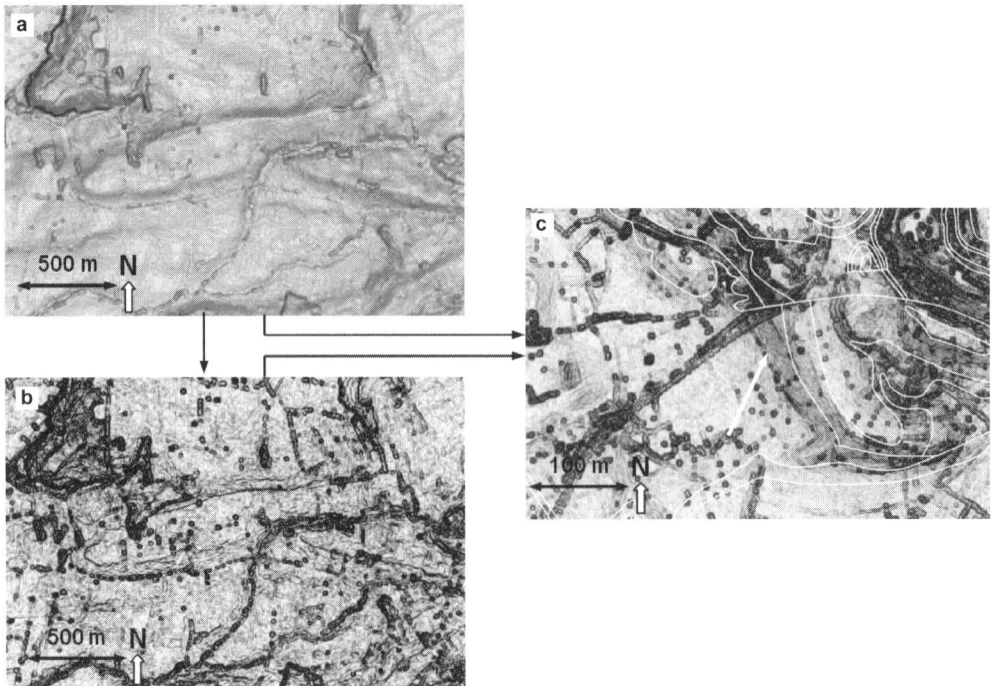

Fig. 7. Methodology for synthetic feature mapping. (**a**) Derived map classifying landscape into areas of similar slope angle, darker shading illustrates steeper slopes. (**b**) Derived map highlighting adjacent areas of dissimilar slope angle, that is, 'break of slope', darker shading illustrates break in slope. (**c**) Slope map and classified slope map combined with existing linework; discrepancy of the Penarth Group shown by the arrow; existing linework in white. Elevation data from Intermap Technologies.

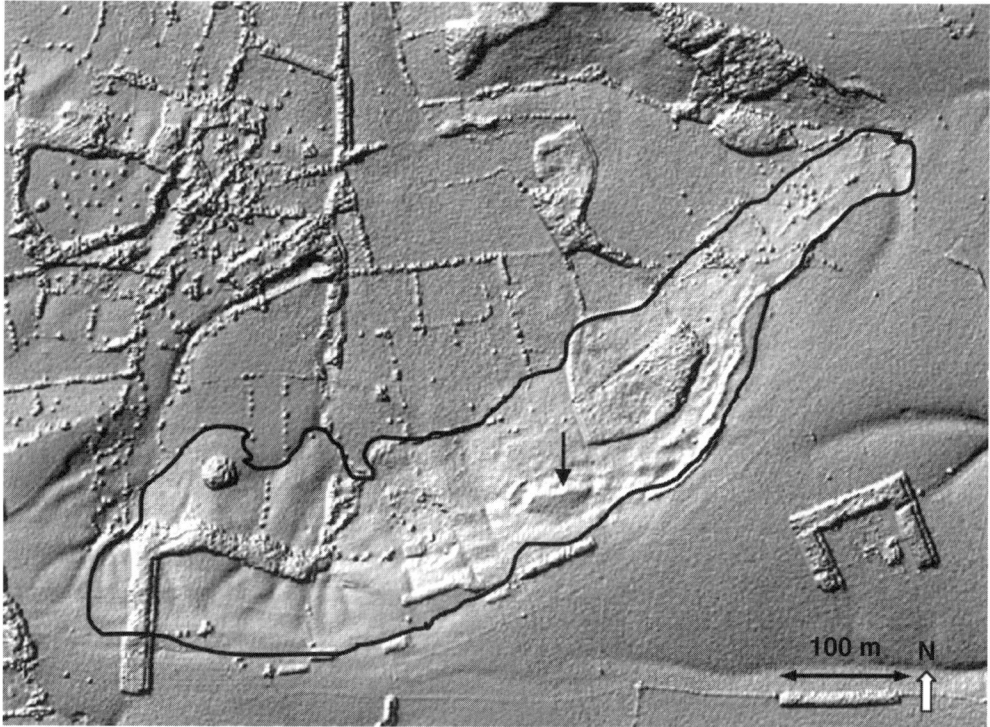

Fig. 8. Area of landslides near Oxpasture Wood, shown by the textural variation within the NEXTMap shaded relief DSM. The black line illustrates the probable extent of mass movement; the arrow shows the location of the rotated block. Elevation data from Intermap Technologies.

In order to provide a rapid initial appraisal of the existing linework, techniques were developed to automatically extract the main landscape features, as described by Ford (2007). This 'synthetic feature mapping' process was carried out with ArcMap using the Spatial Analyst extension; the method is outlined in Figure 7. First, the DSM was classified by slope angle (Fig. 7a). The slope angle map was then classified; this produced the second derivative of the DSM and highlighted areas of dissimilar slope angle, effectively showing breaks of slope (Fig. 7b). Symmetrical crest and valley features were defined by an appraisal of slope facing direction. By combining these derived maps and overlaying existing geological linework, it is possible to identify discrepancies between landscape features shown on the DSM and existing mapping (Fig. 7c). Figure 7c shows the Penarth group following closely the relatively steeper sloping ground illustrated by the darker shading. However, as the unit follows the hillside northwards, the unit is shown to be no longer constrained by the steeper ground (as shown by white arrow) and moves up the slope to the flatter ground at the top of the hill. This appears to be wrong and it is more likely that

this unit would remain on the steeper ground. Detailed contour and elevation information were not available at the time of the original survey and would make a contribution to the inaccuracies identified. Positioning geological boundaries would be difficult in areas where there was little outcrop and subtle landform features. Cultural and vegetation features are also brought out by the synthetic feature mapping approach, as the DSM was used, but it was found to be relatively straightforward to separate these from the landscape features by consulting topographical maps and aerial photography. Results using the DTM were not as favourable as using the DSM because the filtering approach used to generate the DSM from the DTM appears to 'erode' some of the subtle topographical features.

Landslides exist on a number of the steeper slopes formed by the Penarth Group, Redcar Mudstone Formation, Ampthill and Oxford clay formations, and the Cretaceous chalks. The existing geological map for York does not show the location or extent of these landslides because they were not mapped during the 1880s survey, being noted with a comment only on the field maps. With an increasing awareness of geological hazards, the accurate

Fig. 9. Landslides near Oxpasture Wood, the large landslip block on the left is arrowed in Figure 8.

mapping and understanding of mass movement deposits has become increasingly important. The NEXTMap DSM displayed a marked contrast in surface texture between bedrock areas subject to landslide activity and areas with more stable slopes. The most effective processing techniques were the generation of shaded relief and slope images. An example of the undulating texture caused by landsliding can be seen in the shaded relief image in Figure 8. The resolution of

Fig. 10. NEXTMap manipulation in the field using a ruggidized tablet PC with integrated GPS, running customized ArcMap Software. Elevation data from Intermap Technologies.

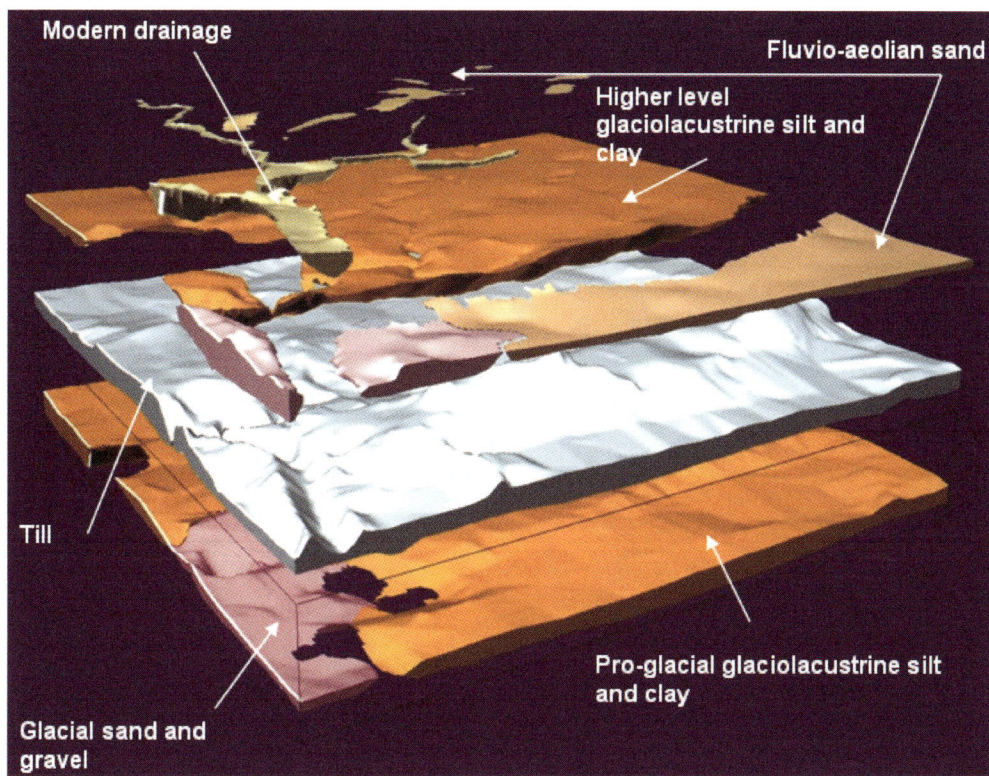

Fig. 11. Three-dimensional 'exploded' geological model of the superficial deposits around York, with surface elevation information from the NEXTMap DTM. Elevation data from Intermap Technologies.

NEXTMap is high enough to identify individual rotated blocks within the landslide complex (Figs 8 & 9). In this way it is possible to not only identify areas of possible landsliding, but to begin to understand the failure mechanisms responsible for the movement.

NEXTMap manipulation in the field and 3D modelling using NEXTMap

A digital data capture system is currently being introduced within the BGS as a replacement for paper fieldslips and notebook as a component part of the move towards a complete digital mapping workflow (Jordan *et al.* 2005). The system is based on a customized version of ArcMap running on a ruggedized tablet PC with an integrated GPS, allowing the capture of text, polygons and lines, and form-based data (Fig. 10). One significant benefit of the system is the ability to manipulate NEXTMap data on a site-specific basis, such as adjusting artificial shading and colour ramping parameters, according to local topographical variations. The system also includes a 'structure contour tool' for calculating where a geological bed will crop out based on structural measurements. Given the location of an observation point where the angle and direction of dip of the bed are known, and assuming the dip is uniform, the tool can calculate an equation for the plane of the bed. Where this plan intersects with the ground surface it can be predicted where the bed will crop out. The NEXTMap DTM is used by the tool to provide elevation data for the calculation and allows the calculation to be completed in the field.

The BGS is increasingly producing 3D geological models for geological analysis and for external clients, as described by Kessler *et al.* (2005). The terrain surface is a key input to a 3D geological model, providing the top surface of the geological sequence and an elevation from which to hang boreholes. To assist in the geological mapping of the York sheet a geological model was constructed to investigate the superficial deposits in York city centre, allowing an appraisal of a large number of boreholes. NEXTMap data provided the surface constraint for the constructed units. The resultant model is shown in Figure 11, with the individual units vertically separated.

Conclusions

The NEXTMap Britain elevation model has provided many opportunities to enhance the geological mapping of bedrock, structure, and superficial and artificial deposits, and has been successfully integrated into the digital and conventional mapping workflows of the Vale of York mapping team.

A variety of techniques have been applied throughout the mapping process and NEXTMap interpretation has made an important contribution to the understanding of the extensive glacial and proglacial deposits found in the Vale of York.

NEXTMap has been used to identify landform features that relate to the underlying geology, providing an overview of the landscape and reducing the time needed to map out features in the field. Techniques have also been developed to automatically generate these landform features in areas of well featured bedrock.

The dataset has also been used to identify areas where landsliding has occurred, for accurate mapping of artificial ground and as a key surface for input into 3D geological models.

Thanks are given to L. Austin, S. Egan, K. McManus and C. Foster for reviewing the manuscript and making constructive comments that have improved the work. The paper is published with permission of the Executive Director, British Geological Survey, Natural Environment Research Council.

References

BONHAM-CARTER, G. F. 1994. *GIS for Geoscientists.* Elsevier Science, Oxford.

BRADWELL, T., STOKER, M. & KRABBENDAM, M. 2007. Megagrooves and streamlined bedrock in NW Scotland: the role of ice streams in landscape evolution. *Geomorphology*, **97**, 135–136, doi: 10.1016/j.geomorph.2007.02.040.

COOPER, A. H. & BURGESS, I. C. 1993. *Geology of the Country around Harrogate.* Memoir of the British Geological Survey, Sheet 62. British Geological Survey, England and Wales.

COOPER, A. H. & GIBSON, A. 2003. *Geology of the Leeds District – A Brief Explanation of the Geological Map. Sheet Explanation of the British Geological Survey, 1:50 000 Sheet 70.* British Geological Survey, England and Wales.

FORD, J. 2007. *Synthetic Feature Mapping – Derivation of Feature Mapping Elements from Digital Elevation Models.* British Geological Survey Open Report, **OR/07/030**. British Geological Survey. (In press).

FOX-STRANGWAYS, C. 1884. *The Geology of the Country North-east of York and South of Malton (Explanation of Quarter Sheet 93NE, New Series Sheet 63).* Memoir of the Geological Survey. British Geological Survey, England and Wales.

GAUNT, G. D. 1981. Quaternary history of the southern part of the Vale of York. *In*: NEALE, J. & FLENLEY, J. (eds) *The Quaternary in Britain.* Pergamon Press, Oxford, 82–97.

GAUNT, G. D. 1994. *Geology of the Country around Goole, Doncaster and the Isle of Axholme.* Memoir of the British Geological Survey, Sheets 79 and 88. British Geological Survey, England and Wales.

HALL, M., HOWARD, A. S., ASPDEN, J. A., ADDISON, R. & JORDAN, C. J. 2004. *The Use of Anaglyph Images for Geological Feature Mapping.* British Geological Survey Internal Report, **IR/04/004**. British Geological Survey, England and Wales.

INTERMAP TECHNOLOGIES 2007. *Intermap Product Handbook and Quick Start Guide.* Standard edn, Version 4.0. Intermap Technologies, Denver, CO.

JORDAN, C. J., BEE, E. J., SMITH, N. A., LAWLEY, R. S., FORD, J. R., HOWARD, A. S. & LAXTON, J. L. 2005. The development of digital field data collection systems to fulfil the British Geological Survey mapping requirements. *In*: CHENG, Q. & BONHAM-CARTER, G. (eds) *GIS and Spatial Analysis: Proceedings of IAMG '05: The Annual Conference of the International Association for Mathematical Geology, Toronto, August 21–25, 2005.* International Association for Mathematical Geology, Kingston, Ontario, 886–891.

KESSLER, H., LELLIOTT, M. *ET AL.* 2005. 3D geoscience models and their delivery to customers. *In*: RUSSELL, H., RICHARD, C., BERG, L. & THORLEIFSON, H. (convenors) *Three-Dimensional Geologic Mapping for Groundwater Applications: Workshop Extended Abstracts: Salt Lake City, Utah, 15 October 2005.* Geological Survey of Canada, Ottawa, Ontario, 39–42.

ONORATI, G., VENTURA, R., POSCOLIERI, M., CHIARINI, V. & CRUCILLÀ, U. 1992. The digital elevation model of Italy for geomorphology and structural geology. *Catena*, **19**, 147–178.

POWELL, J. H., COOPER, A. H. & BENFIELD, A. C. 1992. *Geology of the Country Around Thirsk.* Memoir for 1:50 000 Geological Sheet 52. British Geological Survey, England and Wales.

PIKE, R. 1992. Machine visualisation of synoptic topography by digital image processing. *US Geological Survey Bulletin*, **2016**, B1–B12.

RAWSON, P. F. & WRIGHT, J. K. 1995. Jurassic of the Cleveland Basin. *In*: TAYLOR, P. D. (ed.) *Field Geology of the British Jurassic.* Geological Society, London, 173–208.

SMITH, M. J. & CLARK, C. D. 2005. Methods for the visualisation of digital elevation models for landform mapping. *Earth Surface Processes and Landforms*, **30**, 885–900.

SMITH, M. J., ROSE, J. & BOOTH, S. 2006. Geomorphological mapping of glacial landforms from remotely sensed data: an evaluation of the principal data sources and an assessment of their quality. *Geomorphology*, **76**, 148–165.

Selecting the appropriate digital terrain model: an example from a hazard mapping exercise

H. K. RUTTER[1], R. NEWSHAM[2], D. G. MORRIS[3] & A. A. McKENZIE[1]*

[1]*British Geological Survey, Maclean Building, Crowmarsh Gifford, Wallingford, Oxfordshire OX10 8BB, UK*

[2]*British Geological Survey, Kingsley Dunham Centre, Keyworth, Nottingham NG12 5GG, UK*

[3]*Centre for Ecology and Hydrology, Maclean Building, Crowmarsh Gifford, Wallingford, Oxfordshire OX10 8BB, UK*

Corresponding author (e-mail: aam@bgs.ac.uk)

Abstract: In the UK national derived geological datasets are increasingly being produced, many of which are based on NEXTMap digital terrain model (DTM) or digital surface model (DSM) data. These include groundwater level and land stability datasets. Any DTM is a model of the land surface and under different conditions may have differing degrees of accuracy. This paper compares the NEXTMap data, derived from airborne Interferometric Synthetic Aperture Radar (IfSAR) data, with other frequently used datasets derived from contours and point data; in particular, the Integrated Hydrological Digital Terrain Model (IHDTM), a terrain model that was originally derived from Ordnance Survey (OS) 1:50 000 scale contours, and a DTM interpolated from Land-Form PROFILE data. This initial comparison of the DTMs has highlighted some issues with the NEXTMap data: first, that of elevation inaccuracy in woodland areas; and, secondly, the shadowing effect caused by the side-looking scanner. It also highlights the problems of using DTMs created from contour data in areas of low relief. The development of an uncertainty layer would enable a user to decide whether the DTM was appropriate in certain areas, and could also be incorporated into uncertainty models for the derived national datasets.

Increasingly, national digital datasets of geo-hazards, covering England, Wales, Scotland and Northern Ireland, are being generated within the British Geological Survey (BGS); many of which use a digital terrain model (DTM) in their development. Examples include national groundwater level, ground stability and soluble rocks datasets. These datasets are aimed at helping consumers, including homebuyers and the insurance industry, to make informed decisions about the geological and hydro-geological hazards of a particular site. An accurate DTM of the entire country, from which both height and slope information could be extracted, is required, and the NEXTMap Interferometric Synthetic Aperture Radar (IfSAR) DTM produced by Intermap Technologies was purchased for this purpose.

The initial measurement produced from any IfSAR is a digital surface model (DSM) that includes not only natural terrain but also other radar reflective objects, including buildings and vegetation. A modelled DTM is produced from the DSM by filtering the radar signal to remove buildings and vegetation. This algorithm may be ineffective in areas of dense vegetation, where volume scattering of the radar signal may prevent any returns from the true ground surface, and large stands of trees have been shown to lead to inaccuracies in the DTM (Dowman *et al.* 2003), which causes problems when using these data. For example, when assessing ground stability, slope is important. At the edge of large stands of trees, the elevation changes rapidly from the 'true' elevation of the non-wooded ground to the height of the tree canopy, thus spuriously predicting high-angle slopes and, potentially, areas of 'highly unstable' ground.

A short study was carried out to assess the differences between the NEXTMap DTM and another national DTM, the Integrated Hydrological Digital Terrain Model (IHDTM), with two aims: (i) to assess the effects of the known limitations of the DTM in areas of dense vegetation on the resulting models; and (ii) to assess whether a difference map between the two datasets could be used to identify areas of woodland, in order to remove these from the NEXTMap data.

The study was primarily concerned with establishing the performance of DTMs that had already been used to produce national scale geohazard datasets. The study did not, therefore, consider other data sources; for instance, DSMs derived from Light Detection And Ranging (LIDAR) that are commonly use for detailed hazard assessment in local areas, and are increasingly available at a national scale.

From: FLEMING, C., MARSH, S. H. & GILES, J. R. A. (eds) *Elevation Models for Geoscience.*
Geological Society, London, Special Publications, **345**, 67–74.
DOI: 10.1144/SP345.8   0305-8719/10/$15.00 © The Geological Society of London 2010.

This paper briefly describes the datasets, compares them and highlights some of the issues that should be considered when using the data. It focuses on the fact that the two different types of DTM have different strengths and weaknesses. Although much of the following work was carried out using a national map of the differences between the NEXTMap DTM and the IHDTM, more detailed investigation took place in the Slough area. For this purpose, a DTM derived from Land-Form PROFILE (Ordnance Survey (OS) 1:10 000) data was used.

Datasets investigated

Two types of DTM that have been used for national geohazard assessment are described and reviewed:

- DTMs created from IfSAR data, in particular Intermap's NEXTMap Britain: 5 m grid.
- DTMs created by the interpolation of line and point data, in particular the IHDTM produced by the Centre for Ecology and Hydrology (CEH): 50 m grid.

NEXTMap DTM

NEXTMap uses a technique called IfSAR – Interferometric Synthetic Aperture Radar. An airborne radar sensor is used to produce a terrain model by using phase differences between separated receivers, coupled with accurate GPS information on the sensors position to derive the height and x, y position of the target. The programme to collect these data in the UK is described in Mercer (2007). This approach provides a rapid and accurate way of collecting height data. The raw data can be used as a DSM. The data are also filtered using a bald-earth algorithm (Wang *et al.* 2001) that iteratively filters the DSM to remove isolated topographical high points, which are assumed to represent buildings or vegetation, to try to preserve topographical features to produce a DTM. Further information is given in the *Intermap Product Handbook* (Intermap Technologies 2007). The unfiltered DSM provides high detail, revealing some subtle topographical features. In particular, it provides good detail in areas of low relief. In the DTM subtle topographical features are less distinct (Smith *et al.* 2006). Issues concerning the accuracy of the DTM and DSM are discussed later. The NEXTmap dataset for the England, Wales and southern Scotland was collected in 2002.

IHDTM (derived from OS Land-Form PANORAMA and 1:50 000 rivers)

The IHDTM was created by the Centre of Ecology and Hydrology as a hydrologically validated DTM by digitizing OS Land-Form PANORAMA contours from the Landranger 1:50 000 scale map series. The vertical interval of the contours is 10 m. In general, the contours are continuous features across a 20×20 km tile, except in certain circumstances:

- in areas with steep slopes where some contours may have been omitted.
- where they are coincident with man-made features.
- where they are coincident with active quarries, spoil heaps, gravel pits or open-cast mines.

Additional information is provided by spot heights on some hilltops, heighted lake shores and the high water line. Further information on the Land-Form PANORAMA product is available from the Ordnance Survey (Ordnance Survey 2004). The height data were complemented by a digitized river network (1:50 000) that was used to control the interpolation in the base of valleys and to ensure that flow directions derived from the DTM would be accurate (Morris & Flavin 1990).

Land-Form PROFILE DTM

For an area close to Slough, a 5 m DTM was interpolated from OS Land-Form PROFILE contours and spot heights (from OS's 1:10 000 scale mapping) plus manually inserted control points (Ordnance Survey 2001). The controls had to be added in areas with low relief to control the interpolation of the surface. Contour interpolation was selected for compatibility with geological mapping that had been prepared using the same base map.

Vertical and horizontal resolution

The vertical and horizontal resolution of the datasets are summarized in Table 1. However, as described below, there are issues with both datasets, and these figures should only be used together with an understanding of these issues.

The vertical accuracy of the NEXTMap DSM and DTM is quoted in statistical terms. However, the conditions under which these specifications apply must be considered. The specifications represent upper limits to the vertical accuracy when tested on unobstructed, moderately sloped terrain. Particular terrain and terrain-cover situations may lead to larger errors. In particular:

- shadow effects behind, and layover effects in front of, tall buildings and other tall features.
- the filtering will not remove features greater than around 100 m in all directions; that is, groups of buildings or large stands of trees. In such areas the DTM points are above ground, and there are 'edge effects' around the boundaries.

Table 1. *Horizontal and vertical accuracy of the NEXTMap and OS-based digital datasets*

Model	Horizontal resolution (m)	Stated vertical accuracy	Reference
NEXTMap DSM (product types I and II)	5	0.5–1 m RMSE	Intermap Technologies (2007)
NEXTMap DTM (product types I and II)	5	0.7–1 m RMSE	Intermap Technologies (2007)
IHDTM (derived from OS Land-Form PANORAMA 10 m contours, lake shoes and coastline, plus 1:50 000 rivers)	50	'Generally better than 3 m' (contours) 5 m (DTM) ('Typically better than half of the contour interval')	Morris & Flavin (1990)
OS Land-Form PROFILE	10	1–1.8 m (contours) 2.5–5 m (DTM)	Ordnance Survey (2001)

- rapid changes of terrain, such as at road embankments, may not be preserved owing to inadequate sampling density.
- Slopes greater than 10° cause reduced accuracy.

For the IHDTM, the ability of the interpolated DTMs to represent ground features depends largely on the density of the original height data and on the nature of the terrain. Thus, in flat areas with little height information, contours and spot height data may be sparse, and the interpolation can create irregularities in the DTM.

Results

A map was created to show the difference between the NEXTMap data and the IHDTM for the UK. It was hoped that this difference map might help in understanding the extent of how the NEXTMap DTM behaved in wooded areas. It also identified other differences, leading to further investigation of the discrepancies.

Overall, differences of up to 5 m are frequent, making up just over 10% of the UK. A proportion of these large differences may be partly due to the dissimilar scales: when the NEXTMap data were subtracted from the IHDTM, the NEXTMap data had to be generalized by taking the average of the 100 NEXTMap cells that are represented by each cell of the IHDTM. In areas with rapidly varying topography, there is the potential to introduce errors. The largest source of difference, as was expected, appears to be woodland areas. The NEXTMap DSM records the approximate height of the tree stand; this effect is only filtered out for small stands of trees in the DTM. Such woodland areas can easily be outlined in the difference map in an area of upland with extensive forested plantations (see Fig. 1).

In a small area at Burnham Beeches, near Slough, UK, the differences were investigated more comprehensively. In this area the OS Land-Form PROFILE (1:10 000) contour data were used together with manually inserted control points to create a 5 m DTM; the purpose was to be able to compare the OS and NEXTMap data on a similar scale. The Land-Form PROFILE-based DTM was first compared with the IHDTM, with which there was good correspondence despite the difference in resolution (see Fig. 2). The 'stepped' effect that can be seen in the IHDTM data is a result of re-gridding the 50 m resolution data to a 10 m grid for comparison with the other datasets. When compared with the NEXTMap DTM, there were major differences in height – up to a maximum of 26 m (Fig. 2). Detailed examination of the data suggested that the OS data were usually close to the elevations recorded at independently levelled control points (boreholes). In Figure 2 the major differences in the west were across a wooded valley, where it appears that the NEXTMap data failed to record the presence of the valley. In this area there are likely to be two causes for the anomaly: first, the woodland effect; and, secondly, the shadowing effect of the side-looking radar. The foreshortening and shadowing of side-looking airborne data are likely also to be problems in urban and mountainous areas. Further to the east, where there was a stream cutting through the woodland, the NEXTMap data were better controlled, and are closer to the values obtained from the OS data; while the IHDTM, because of its coarse resolution, missed this feature.

The differences were also examined by summarizing them by land cover class, using the CEH Land Cover Map 2000 (LCM 2000) data. LCM 2000 is derived from a computer classification of satellite scenes, obtained mainly from Landsat satellites, and subdivides the land cover into 27 broad habitat subclasses (not all of which were represented in this area) (Fuller *et al.* 2001). Differences were

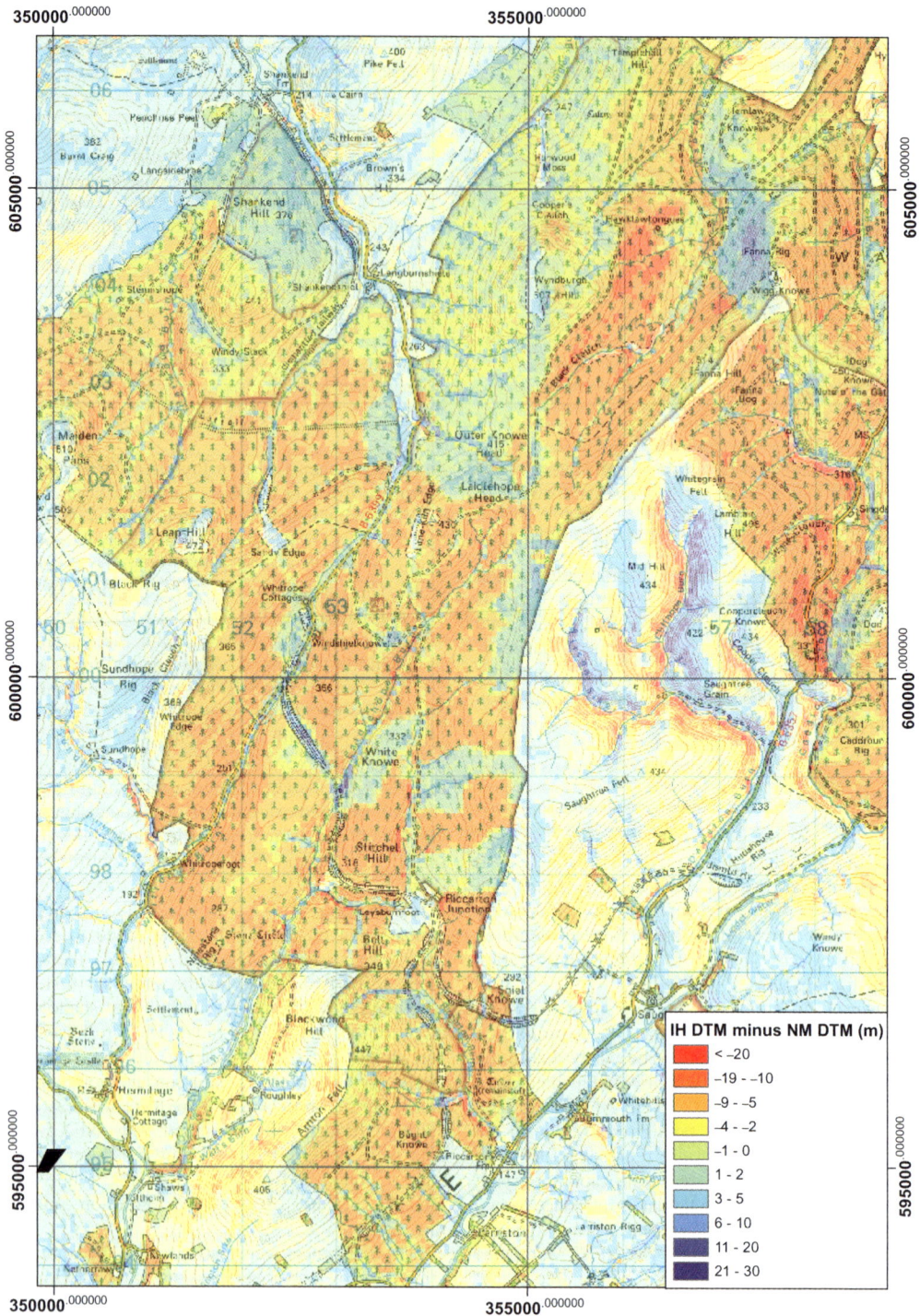

Fig. 1. Illustration of the elevation difference in woodland areas. It should be noted that the elevation difference is not uniform: areas where the difference is minimal might be areas that have been cleared or where there is new plantation. This map is reproduced from Ordnance Survey topographical material with the permission of the Ordnance Survey on behalf of The Controller of Her Majesty's Stationery Office, ã Crown copyright. Unauthorized reproduction infringes Crown Copyright and may lead to prosecution or civil proceedings. Licence Number: 100017897 [2008].

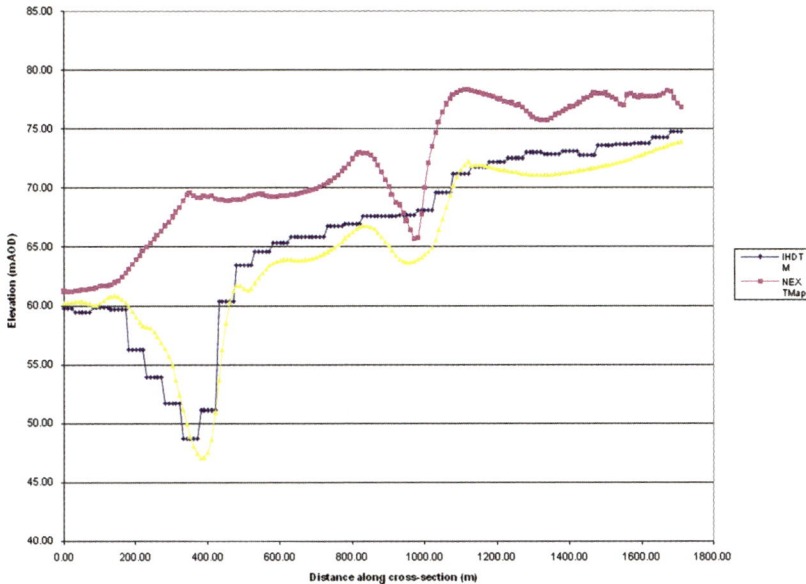

Fig. 2. Cross-section through the Burnham Beeches area illustrating the difference in elevation between the IHDTM, NEXTMap DTM and DTM created from Land-Form PROFILE data. Note that the 'stepped' effect that can be seen in the IHDTM is a result of sampling the 50 m resolution data every 10 m.

summarized within each of the land cover class polygons (see Table 2).

Table 2 reveals that the principal differences in elevation are for woodland areas, and also that the greatest standard deviation is for broad-leaved woodland. That is, although the mean difference for broad-leaved woodland is −6 m, the standard deviation of 5.27 m shows there is also considerable variability in the differences within this land cover class. This may be partly due to the shadowing effects observed in this area.

The variability of the differences within woodland classes may be important when formulating

Table 2. *Elevation difference (metres) (OS Land-Form PROFILE minus NEXTMAP DTM) summarized by land cover class*

Class		Mean	STD
1.1	Broad-leaved woodland	−6.00	5.27
2.1	Coniferous woodland	−4.58	1.97
16.1	Inland bare ground	−2.47	1.95
17.1	Suburban	−0.71	1.59
4.3	Non-rotational horticulture	−0.71	0.15
5.1	Improved grassland	−0.62	1.81
4.1	Arable cereals	−0.59	1.27
6.1	Neutral grassland	−0.49	1.43
4.2	Arable horticulture	−0.48	1.72
17.2	Continuous urban	−0.37	2.48

an approach to correct the DTM in such areas, and merits further investigation. It does indicate that land cover mapping might be used to calculate a national uncertainty map for the NEXTMap DTM. The uncertainty map could be used in calculating the uncertainty of datasets derived from the DTM, and to prioritize corrections.

Other issues that may lead to differences between the DTMs include the following.

Land surface change

Recent land use may change the topography but these changes may not be recorded (contours are not always recorded in active quarries, spoil heaps, gravel pits or open-cast mines) or may be more recent than the data used to create the IHDTM. The issue of land surface change is of importance to all datasets in areas that are actively quarried and filled: each dataset may record any one of the natural land surface, the quarried surface or the backfilled surface, or may record a mixture of surfaces. In some cases, it may be important to track the date that data were collected – there is a minimum of 15 years difference in the date of the IHDTM and the NEXTMap DTM.

Horizontal accuracy

The horizontal positioning of data is of critical importance in areas with high topographical relief.

The initial investigation of the difference map showed that, in some cases, the original OS Land-Form PANORAMA contours were offset across the published map boundaries. In hilly areas a small offset in *x* or *y* can create a large change in elevation. Although this can result in an offset in the height of the IHDTM across the map boundary, it may not be noticeable; however, it becomes apparent in the difference map (Fig. 3).

Areas with limited height data

Any areas where there are few contours or spot height data will have implications for the reliability of the IHDTM, as there are few data to control the interpolation.

Along the coast, as there are no height data, an arbitrary value of 3 m was assigned to the mean high water mark in the interpolation of the IHDTM. This is probably realistic in many areas. However, there are some areas where a zero contour slightly inland causes the DTM to show lower heights inland than at the coast.

The lack of contour data can cause problems across some quite extensive areas, the area around the Wash being a particular anomaly (Fig. 4). This area has very few contours, and the difference values suggest that the IHDTM may be several metres too high. In this area the differences close to the coast are small, with the IHDTM deviating from the NEXTMap data inland, where there are no data.

Conclusions

The initial driver for this investigation was the problem caused by the change in elevation across woodland areas with the NEXTMap DTM data; in particular, the spurious prediction of steep slopes and potential slope stability problems at the edges of woodland. The BGS are currently investigating approaches to correct the DTM in such areas. Without correction, the use of this dataset in the derivation of national geohazard datasets may be constrained.

All DTMs will have issues concerning the vertical accuracy of the elevation values. It is important to understand that quoted accuracies may be for specific conditions, and that within certain areas accuracies may be much lower. Although the datasets discussed in this paper are of different resolutions, the issues identified are generally scale-independent.

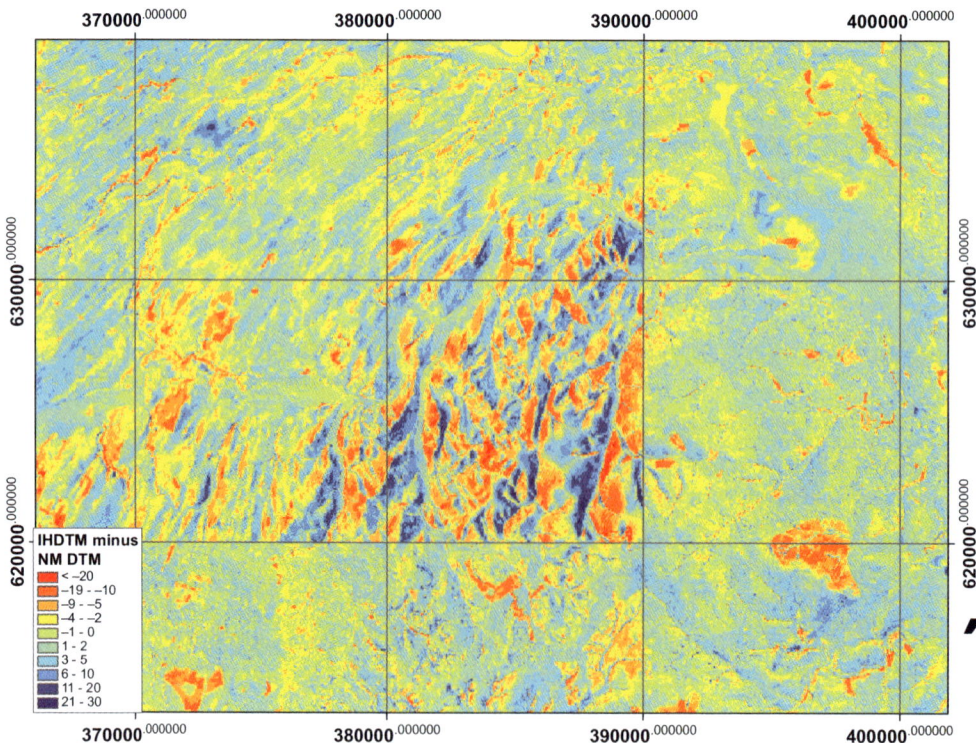

Fig. 3. Anomalies caused in the difference map where there are slight inaccuracies at map-sheet boundaries in the contour data used to create the IHDTM.

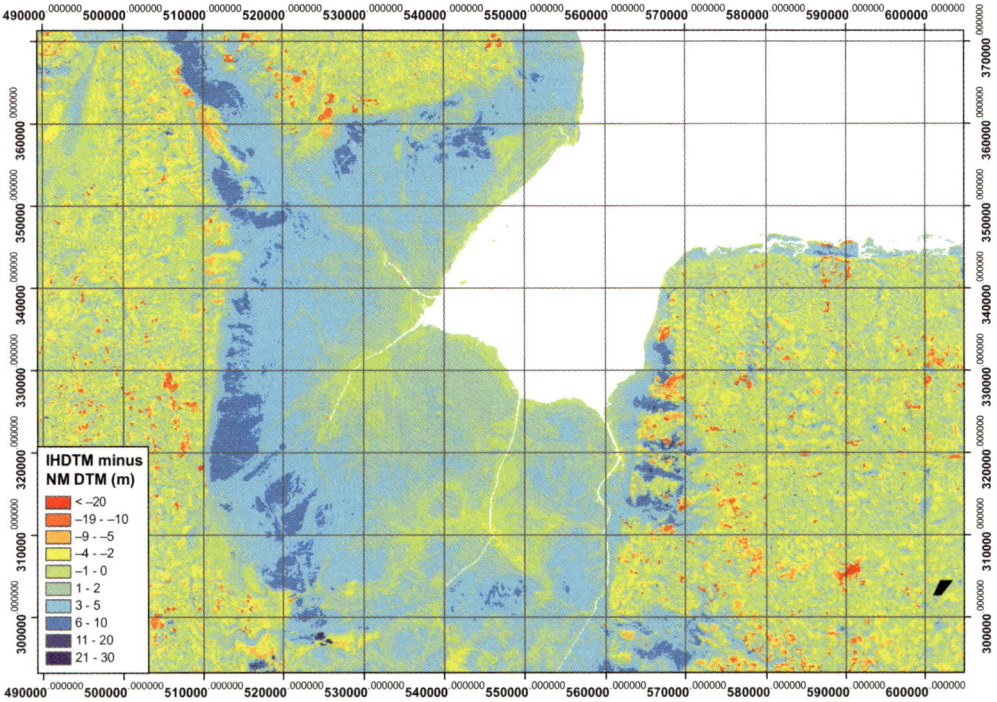

Fig. 4. Effect of the lack of contour and spot height data in areas of low relief.

A comparison of the IHDTM and NEXTMap DTM at a national scale has highlighted the issue of inaccuracies in wooded areas, partly as a result of limitations in the bare earth algorithm used by NEXTMap. Differences between the datasets were correlated with land cover classes, showing clearly the role played by woodland, but also pointing to differences between the behaviour of broad leaf and coniferous land cover.

The work carried out also identified issues with the IHDTM. These issues included: errors in the registration of contours across map sheets, which leads to systematic errors in areas of high relief, and a paucity of data in areas of low relief, especially in coastal areas.

By looking at an area in greater detail the effect of woodland and shadowing or foreshortening owing to the side-looking instrumentation on the NEXTMap data can be compared to issues arising from the low resolution of the IHDTM.

While further work will be required to fully understand the limitations of these DTM, and to understand how the NEXTMap DTM performance may be increased in wooded areas, the current study has indicated how an uncertainty layer might be developed enabling the user to choose a DTM that is fit for purpose, and could also be incorporated into uncertainty models for the derived national datasets.

References

DOWMAN, I., BALAN, P., RENNER, K. & FISCHER, P. 2003. An evaluation of NEXTmap terrain data in the context of UK National Datasets. Report to Getmapping. Available online at: http://www.intermap.com/uploads/1170362281.pdf.

FULLER, R. M., SMITH, G. M., SANDERSON, J. M., HILL, R. A. & THOMSON, A. G. 2001. The UK Land Cover Map 2000: construction of a parcel-based vector map from satellite images. *Cartographic Journal*, **39**, 15–25.

INTERMAP TECHNOLOGIES 2007. *Product Handbook and Quick Start Guide*, Version 4.2. Intermap Technologies Ltd. Intermap Technologies, Denver, CO.

MERCER, B. 2007. National and regional scale DEMs created from airborne INSAR. *In*: STILLA, U. *ET AL.* (eds) PIA07. *International Archives of Photogrammetry, Remote Sensing and Spatial Information Sciences*. International Society for Photogrammetry and Remote Sensing (ISPRS), Beijing, **36**(3/W49B), 113–117. Available online at: http://www.intermap.com/uploads/1191434770.pdf.

MORRIS, D. G. & FLAVIN, R. W. 1990. A digital terrain model for hydrology. *Proceedings 4th International Symposium on Spatial Data Handling, Zurich*, **1**, 250–262.

ORDNANCE SURVEY 2001. *Land-Form Profile User Guide*. Available online at: http://www.ordnance-survey.co.uk/oswebsite/products/landformprofile/pdf/profil_w.pdf.

ORDNANCE SURVEY 2004. *Land-Form Panorama User Guide*. Available online at: http://www.ordnancesurvey.co.uk/oswebsite/products/landform-panorama/pdf/Land-Form_PANORAMA_user_guide.pdf.

SMITH, M. J., ROSE, J. & BOOTH, S. 2006. Geomorphological mapping of glacial landforms from remotely sensed data: an evaluation of the principal data sources and an assessment of their quality. *Geomorphology*, **76**, 148–165.

WANG, Y., MERCER, B., TAO, V. C., SHARMA, J. & CRAWFORD, S. 2001. Automatic generation of bald earth digital elevation models from digital surface models created using airborne. *In*: *IfSAR – Proceedings of 2001 ASPRS Annual Conference, 2001*. American Society for Photogrammetry & Remote Sensing, Bethesda, MD. Available online at: http://www.intermap.com/uploads/1170700805.pdf.

The use of elevation models to predict areas at risk of groundwater flooding

A. A. McKENZIE[1]*, H. K. RUTTER[1] & A. G. HULBERT[2]

[1]*British Geological Survey, Maclean Building, Crowmarsh Gifford, Wallingford, Oxfordshire OX10 8BB, UK*

[2]*British Geological Survey, Kingsley Dunham Centre, Keyworth, Nottingham NG12 5GG, UK*

*Corresponding author (e-mail: aam@bgs.ac.uk)

Abstract: Groundwater flooding, which occurs when the groundwater table rises in response to exceptional recharge rates either to the ground surface or to a point where subsurface infrastructure is affected, has been recognized as a significant issue with real economic impacts.

A methodology has been developed to produce maps of groundwater flooding susceptibility, using geological and hydrogeological data. While good geological map data are available in digital form for England and Wales, there are much less data on water levels. These levels are usually measured during the construction of water boreholes, and while there is a national groundwater level monitoring network for regulatory purposes, at a national level data are sparse. To assist in developing a comprehensive map of water levels, the British Geological Survey (BGS) has adopted a number of strategies for data interpolation for areas with limited water level data and a surface has been derived from a terrain model by considering interactions between groundwater and surface water in rivers and lakes. When comparing the calculated levels against the available field measurements, a high correlation was found to exist. However, it was considered that in areas where bedrock aquifers dominate, this interpolated surface was probably inaccurate, and so refinements were developed to improve the modelled water levels surfaces.

The resulting groundwater levels have been used to develop maps of areas where shallow groundwater may pose a risk. With potential changes in groundwater recharge postulated as a result of global climate change, identifying areas prone to flooding from groundwater, or areas where groundwater is likely to increase the impact of surface water flooding, is increasingly important.

Groundwater level datasets are one of the primary resources for hydrogeological interpretation, and knowledge of groundwater level is important for many site-specific geoscientific enquiries; the depth at which groundwater is found is required for groundwater flow and groundwater pollution modelling, it affects the movement of pollution from the ground surface into aquifers, it is important in predicting the viability and cost of drilling water abstraction boreholes, and shallow groundwater can damage buildings and lead to flooding.

During the exceptionally wet winter of 2000–2001 groundwater levels were high throughout much of southern England and a series of groundwater flooding events occurred in the region highlighting the economic impact of such flooding, both to infrastructure such as roads and railways and to private properties. To assist in an understanding of groundwater flooding a project was undertaken to develop a series of groundwater flooding susceptibility maps that combine geological information with data on groundwater level to identify areas where groundwater flooding might occur. One of the main requirements of the project was to collate information on groundwater level.

Groundwater levels are measured by observation of the rest water level in wells and boreholes, and regulatory authorities have developed networks of observation points that are sampled regularly. Observation points, however, are relatively sparse, and are generally focused on aquifers used for public water supply. To supplement water level monitoring data in other areas, a combination of approaches is required using both data collated from observations during well and borehole construction and inferences based on topography and geology.

Groundwater flooding

In a recent review of groundwater flooding for the Department for Environment, Food and Rural Affairs (DEFRA 2004*a*) groundwater flooding was defined as 'being the emergence of water at the ground surface through a natural process' and as 'the type of flooding that can be identified as being caused by water originating from beneath the ground surface from permeable strata through a natural process'. The US Geological Survey (USGS 2000) describes groundwater flooding as follows: 'Groundwater flooding occurs in low-lying areas

From: FLEMING, C., MARSH, S. H. & GILES, J. R. A. (eds) *Elevation Models for Geoscience.*
Geological Society, London, Special Publications, **345**, 75–79.
DOI: 10.1144/SP345.9   0305-8719/10/$15.00 © The Geological Society of London 2010.

when the water table rises above the land surface' (DEFRA 2004*b*). These definitions can be expanded to take account of the way in which groundwater feeds many streams or rivers by providing base flow and also feeds springs through 'natural processes', and to take account of the flooding of basements and foundations and tunnelling infrastructure by groundwater.

Groundwater flooding is a relative and not absolute concept depending on:

- the relative height of local groundwater levels above some notional typical high groundwater level.
- the length of time for which the groundwater levels are raised by a given level above some (arbitrary) high groundwater level.
- the degree of social and economic damage (usually expressed as a cost) associated with the flooding event.

The occurrence of groundwater flooding follows prolonged periods of rainfall and/or intense rainfall events. Southern England experienced significant (region-wide) groundwater flooding events in the winters of 1993–1994, 1994–1995, 2000–2001 and, most recently, in 2002–2003. For example, the 2000–2001 groundwater flooding was associated with unusually high levels of rainfall. For an 8 month period, starting in September 2000, rainfall in SE England was 183% of the long-term average rainfall, equivalent to a more than 100 year return period. Even though unusually high levels of rainfall were experienced over much of the country during the winter of 2000–2001, the groundwater flooding was generally restricted to central southern and SE England. Impacts included flooding of properties, surcharging of sewers, and rail and road disruption. It has been estimated that, excluding the cost of disruption to the rail services, this single groundwater flooding incident alone cost of the order of £800 000. A common feature of groundwater flooding events is their relatively long duration compared with surface water or fluvial flooding, so that costs per property of flooding from groundwater are often in excess of costs associated with an equivalent fluvial flooding event.

Estimates (Defra 2004*a*, *b*) are that up to 1.7 million properties in England may be vulnerable to groundwater flooding.

Developing a groundwater flooding susceptibility map

Until recently, groundwater flooding has received little attention from the hydrogeological community in the UK, and while local knowledge of groundwater flooding may be detailed, unfortunately, this information has not necessarily been collated at a national level. This means it is not easy to assess the extent of historic groundwater flooding events. Groundwater flooding has not often been recorded systematically for a variety of reasons, including the relative rarity of events and the masking of groundwater flooding by simultaneous surface water floods. It was thought to be valuable to develop a methodology that could map areas susceptible to groundwater flooding at the national level, with initial work concentrating on England and Wales.

The following three-stage approach was adopted.

(1) Definition of simple, but geologically and hydrogeologically realistic, conceptual models of groundwater flooding processes. These included flooding associated with permeable superficial deposits in hydraulic continuity with rivers and flooding associated with the direct intersection of a bedrock water table with the ground surface.

(2) Identification of areas where geological conditions are such that one of the conceptual models may apply.

(3) Identification of likely maximum groundwater levels in the areas identified, and an assessment of where those water levels are close enough to the surface to make groundwater flooding possible.

Once the conceptual models were developed the second and third stages were implemented in a GIS. For the second stage existing digital geological mapping at a scale of 1:50 000 was available (Jackson & Green 2003). The compilation of the water level data was more complex, and is discussed below.

Groundwater levels

Groundwater levels are normally measured in observation borehole networks operated by regulators and water abstractors. Water levels can be measured manually or by data loggers. The data collected can be used to monitor groundwater fluctuations and can also be used as the basis for groundwater level contour maps. The observation networks available are, however, limited in their geographical extent and data points are widely spaced. Most monitoring is concentrated in areas of high groundwater abstraction, leaving substantial areas with little monitoring. The principal UK network provides long-term data for major aquifers, but leaves many minor aquifers without data (Doorgakant 1996). In localized areas the available observation data have been contoured to produce maps of groundwater level.

Water levels are also measured routinely on the completion of drilling of water wells, and the British Geological Survey holds these data within a digital database for the majority of England and Wales. This dataset contains over 50 000 points, but the data are complicated by the dynamic nature of groundwater, with levels changing during the year and over time in response to changes in climate, land use and abstraction of groundwater. There is thus no guarantee that any particular measurements are fully representative of current or future conditions.

Water levels are also routinely measured in boreholes during civil engineering site investigations, but these data have, in the past, not often been available in digital form.

National groundwater level compilation

The full range of groundwater level data is available for the calculation of the groundwater level surface used within the development of the flood map; however, it was not a trivial task to produce a coherent national surface for England and Wales.

Three types of groundwater level data were available. These were:

- groundwater levels taken from contours on published hydrogeological maps and from other digitized contours.
- point groundwater level data from databases of water well construction.
- groundwater levels inferred from adjusted river base levels.

Where groundwater level contour data were available they were was used to define the groundwater levels for an area. Outside of these areas, point measurements of groundwater level were used where there was sufficient data density (generally at least one point per km^2). However, both of these datasets were only available for the estimation of water levels in bedrock, so for the assessment of water levels outside the coverage of the contoured and interpolated data a methodology based on a digital terrain model was adopted.

Water levels in areas with limited data

To establish a water level in areas without sufficiently dense measurements a conceptual model originally applied to calculating possible water levels in superficial deposits as part of an investigation into groundwater vulnerability in Scotland (Ball *et al.* 2005) was adopted. The conceptual model assumes that groundwater is in hydraulic continuity with surface water in rivers, lakes or the sea, and that the water table will not be lower than a surface interpolated between these surface water bodies (Fig. 1). In practice this will only be true in a limited number of aquifers, and in cases where there is no significant abstraction from the aquifer. Fortunately, the aquifer types to which the model applies are just those that are most likely to be susceptible to groundwater flooding – permeable superficial deposits close to rivers.

The conceptual model was implemented within a GIS. The river network and terrain used was originally based on a 50 m, hydrologically corrected, digital elevation model (DEM) produced by the Centre of Ecology and Hydrology (Morris & Flavin 1990). At a later stage the process was repeated using a higher resolution terrain model derived from radar interferometry by Nextmap. (Colemand & Mercer 2002).

A river base level (RBL) surface was derived by:

(1) addition of a vertical elevation attribute derived from the DEM to the vector river network.
(2) construction of a TIN (triangulated irregular network) from the vector river network and from the coastline.
(3) interpolation of a 50 m-cell size raster surface from the TIN.

The river base level surface was converted to a depth to groundwater by subtracting the surface from the DEM. The final surface has been denoted the river

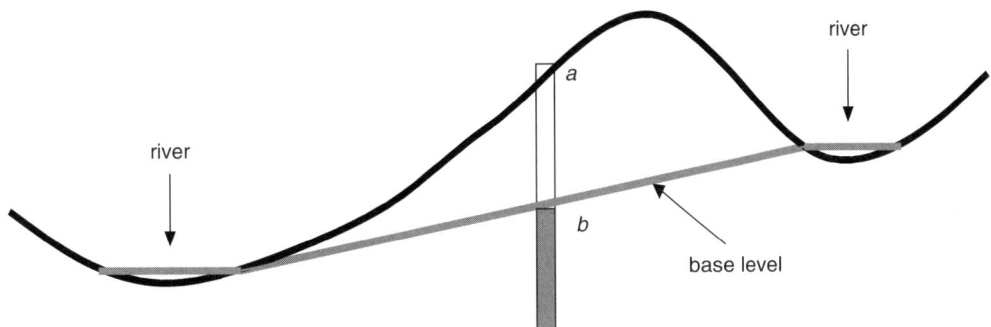

Fig. 1. Cross-section showing the principles used to derive water level. In this cross section the base level has been interpolated between two rivers. A borehole has terrain surface *a* and calculated base level *b*.

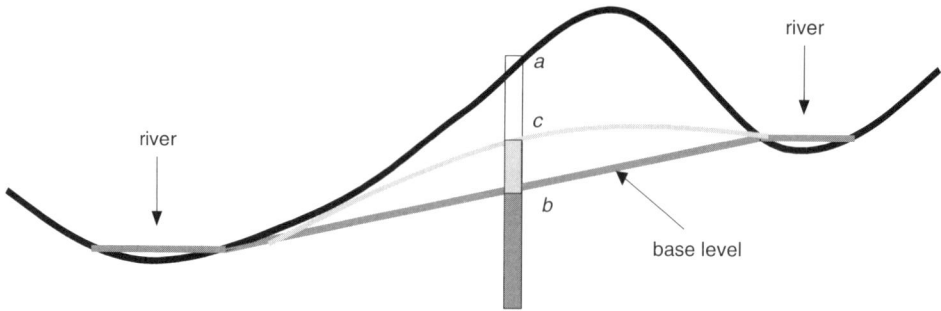

Fig. 2. Cross section showing the principles used to derive adjusted water level. In this cross section the base level has been interpolated between two rivers. A borehole has terrain surface *a*, calculated base level *b* and adjusted base level *c*.

head space (RHS). Assumptions inherent in this process include:

- the level assigned from the DEM reflects the level of water in the river.
- river water level fluctuations are not accounted for.
- there is good hydraulic connection between the river and surrounding permeable superficial deposits.

- no perched aquifers – the RBL would be lower than recorded levels.
- no effects of pumping – the RBL would be higher than recorded levels.

Nevertheless, the surface has been tested against the limited number of direct observations of groundwater level in permeable superficial deposits that are available in digital form, and there is a reasonable correlation between observed and modelled

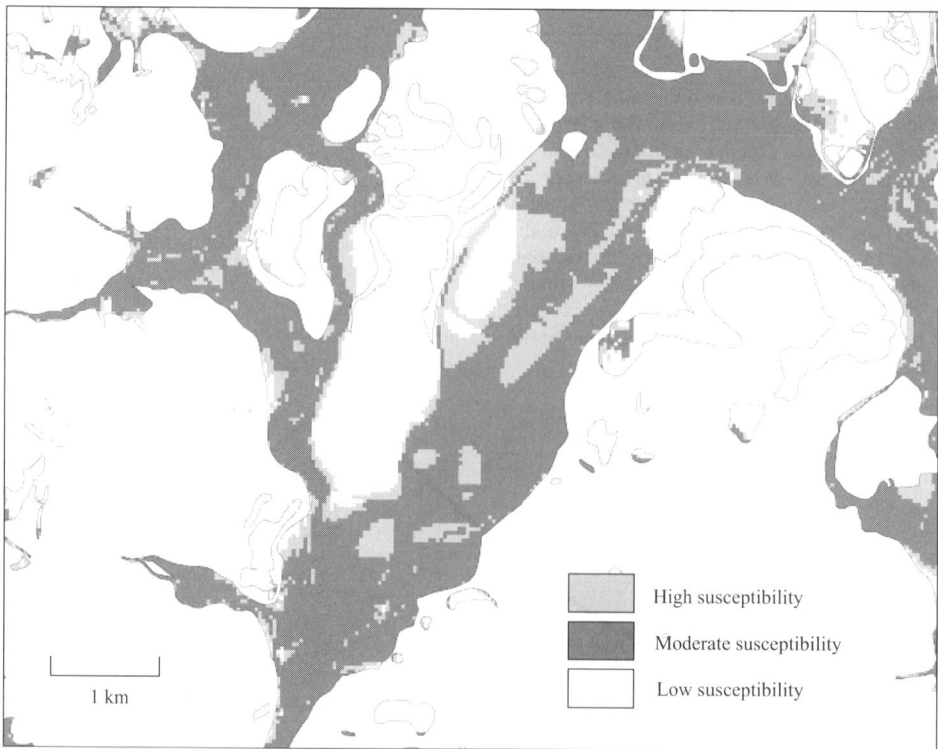

Fig. 3. A groundwater flood susceptibility map produced using the methodology for an area dominated by permeable superficial deposits.

levels. In bedrock aquifers the correlation between observations and the modelled surface is less convincing. This was not unexpected as groundwater levels will generally be higher in the interfluves in response to recharge from precipitation, with water then flowing through the aquifer towards the rivers. The exact form of the water table will be determined by the rate of recharge and the physical properties of the aquifer – its thickness and permeability. Groundwater levels (as metres below ground level) taken from available point observations were regressed against values of RHS. To take some account of variations in permeability a classification of geological units into permeability ranges was used to subdivide the dataset. Only geological formations with high or very high permeability were used, as these were the units where flooding was considered geologically feasible. The following regressions were obtained.

- Very high permeability:
  - water level $= 7.7 + 0.8$ (RHS)
  - $r^2 > 0.6$
- High permeability:
  - water level $= 9.6 + 0.4$ (RHS)
  - $r^2 < 0.2$

Despite the poor correlation coefficients, it was felt that an RHS using the relevant regression coefficients was closer to the real water table than the unadjusted RHS and this was the value used in bedrock aquifers (Fig. 2).

The adjusted RHS could be improved by modelling recharge rates and aquifer properties explicitly, and by developing regression coefficients based on samples more closely constrained to individual geological units. This was not carried out, however, owing to limitations in data availability.

Applying water levels to the calculation of groundwater flooding susceptibility

Once the measured and modelled groundwater level datasets were integrated it was necessary to apply a measure of the likely seasonal variability in levels. For some major aquifers these data were available as contour maps of seasonal fluctuation, and in other aquifers it was derived from time series of groundwater level in observation boreholes, which were used to produce a standard model of fluctuation. The groundwater level, seasonal variability and the terrain model were combined to produce a classification of areas likely to be susceptible to groundwater flooding, ranked according to the closeness of the water table to the land surface. These calculations were limited to the areas already designated as being geologically susceptible. An example is presented as Figure 3.

The resulting susceptibility maps have been compared to the limited information that is available on historical groundwater flooding, and generally provide a satisfactory match. The maps are designed primarily to delineate areas where groundwater flooding should be considered as part of an environmental or site investigation rather than as a tool for site-specific risk assessment.

Conclusion

Groundwater flood susceptibility mapping requires good groundwater level data. Where these data are not available from direct observations then the use of a DEM allowed a modelled groundwater level surface to be constructed at a national scale. While recognizing that the dataset that results has limitations, it allows the estimation of areas susceptible to groundwater flooding to be calculated at a national scale.

With potential changes in groundwater recharge postulated as a result of global climate change, identifying areas prone to flooding from groundwater, or an area where groundwater is likely to increase the impact of surface water flooding, is increasingly important.

References

BALL, D., Ó DOCHARTAIGH, B. É., MACDONALD, A. M., LILLY, A., FITZSIMONS, V., DEL RIO, M. & AUTON, C. A. 2005. Mapping groundwater vulnerability in Scotland: a new approach for the Water Framework Directive. *Scottish Journal of Geology*, **41**, 21–30.

COLEMAND, M. D. & MERCER, J. B. 2002. Nextmap Britain: completing phase 1 of Intermap's global mapping strategy. *GeoInformatics*, **5**, 16–19.

DEFRA 2004a. *Strategy for Flood and Coastal Erosion Risk Management: Groundwater Flooding Scoping Study*. Department for Environment, Food and Rural Affairs Report, LDS23.

DEFRA 2004b. *Making Space for Water. Developing a New Government Strategy for Flood and Coastal Erosion Risk Management in England: A Consultation Exercise*. Department for Environment, Food and Rural Affairs Report, PB 9792.

DOORGAKANT, P. 1996. Groundwater level archive for England and Wales. *In*: GILES, J. R. A. (ed.) *Geological Data Management*. Geological Society, London, Special Publications, **97**, 137–144.

JACKSON, I. & GREEN, C. A. 2003. The digital geological map of Great Britain. *Geoscientist*, **13**, 4–7.

MORRIS, D. G. & FLAVIN, R. W. 1990. A digital terrain model for hydrology. *In*: *Proceedings of the 4th International Symposium on Spatial Data Handling, University of Zurich*. International Geographical Union, 250–262.

USGS 2000. *Ground-Water Flooding in Glacial Terrain of Southern Puget Sound, Washington*. World Wide Web address: http://wa.water.usgs.gov/projects/pugethazards/urbanhaz/GWflooding.htm

Digital elevation models in the marine domain: investigating the offshore tsunami hazard from submarine landslides

DAVID R. TAPPIN

British Geological Survey, Kingsley Dunham Centre, Keyworth,
Nottingham NG12 5GG, UK (e-mail: drta@bgs.ac.uk)

Abstract: Digital elevation models (DEMs) of seabed relief are now commonly available at a number of scales. On a global scale three-dimensional (3D) relief maps of the ocean floor are derived from satellite gravity measurements validated by single-beam echo soundings. On a smaller, more local, scale, the development of multibeam bathymetric mapping technology provides detailed seabed data from which DEMs are derived. Over the past 30 years multibeam bathymetry has replaced single-beam echo soundings as the main tool used to map the sea floor. Multibeam bathymetry has revolutionized our ability to interpret seabed morphology. It has the capability to provide complete seabed coverage and gives a 3D visualization of the seabed not previously available. DEMs derived from multibeam are comparable to those on land. One aspect of the improved seabed visualization is in mapping marine geohazards. Here three DEMs, from Papua New Guinea, Hawaii and the Indian Ocean, are presented. These DEMs have been used to investigate submarine seabed failure and volcanic flank collapse in the context of their tsunami hazard. For these three areas the DEMs contribute to an improved interpretational capability in marine geohazards. In addition, the DEMs underpin newly developed modelling methodologies of landslide-generated tsunami.

Mapping in the marine domain utilizes a number of remote technologies that capitalize on the property of water to transmit sound over great distances. Sound transmission in water is one of the major advantages of morphological and geological mapping at sea over data acquisition on land. Its application to marine mapping has resulted in the development of a variety of remote sensing techniques that include sonars (an acronym for SOund Navigation And Ranging) for mapping both water depth (multibeam) and seabed composition (backscatter intensity). In addition, seismic reflection profiling, that penetrates the seabed, provides images of subseabed structure ranging from shallow levels of sediment thickness up to a penetration depth of several kilometres, which provides data on deep structure. By using a combination of these remote techniques, together with samples obtained from sediment grabs and cores, and deeper penetration from rock coring, a composite three-dimensional (3D) structure of the seabed is obtained.

Of most relevance to this paper are marine technologies that provide water depth and seabed morphology, and that form the basis of digital elevation models (DEMs). The most important of these technologies is multibeam – swath bathymetry that provides data on water depth. A secondary and complementary technology is sidescan sonar that acquires backscatter intensity data of the seabed from which seabed hardness and sedimentary character can be interpreted. The object of this paper, therefore, is to present and discuss DEMs based on multibeam bathymetry (in one instance with associated backscatter intensity data) from three areas: Papua New Guinea (PNG), Hawaii and Sumatra. These data were acquired to specifically investigate the potential hazard from tsunami generated from submarine landslides and volcanic flank collaspe. In addition, the data from PNG formed the basis for the development of new simulations for tsunami from submarine landslides.

Multibeam bathymetry and backscatter intensity

Multibeam uses acoustic signals emitted from a series of transmitters mounted on the hull of a vessel (Fig. 1). Resolution in both horizontal and vertical planes is dependent on both water depth and frequency of the multibeam system deployed. Greater numbers of transmitters and higher frequencies provide more precise bathymetric information. The relationship is complicated but, generally speaking, the deeper the water the lower the frequency of the system used, which results in reduced horizontal and vertical resolution than in shallower waters. For example, a commonly used multibeam system in shallow-water depths of up to 1000 m is the EM1000® (Hughes Clarke *et al.* 1996). The EM1000® operates at a frequency of 95 kHz, producing a fan of 60 beams with $2.4° \times 3.3°$ beam widths, with a total angular swath width of 150°. It is at its best for water

From: FLEMING, C., MARSH, S. H. & GILES, J. R. A. (eds) *Elevation Models for Geoscience*.
Geological Society, London, Special Publications, **345**, 81–101.
DOI: 10.1144/SP345.10   0305-8719/10/$15.00 © The Geological Society of London 2010.

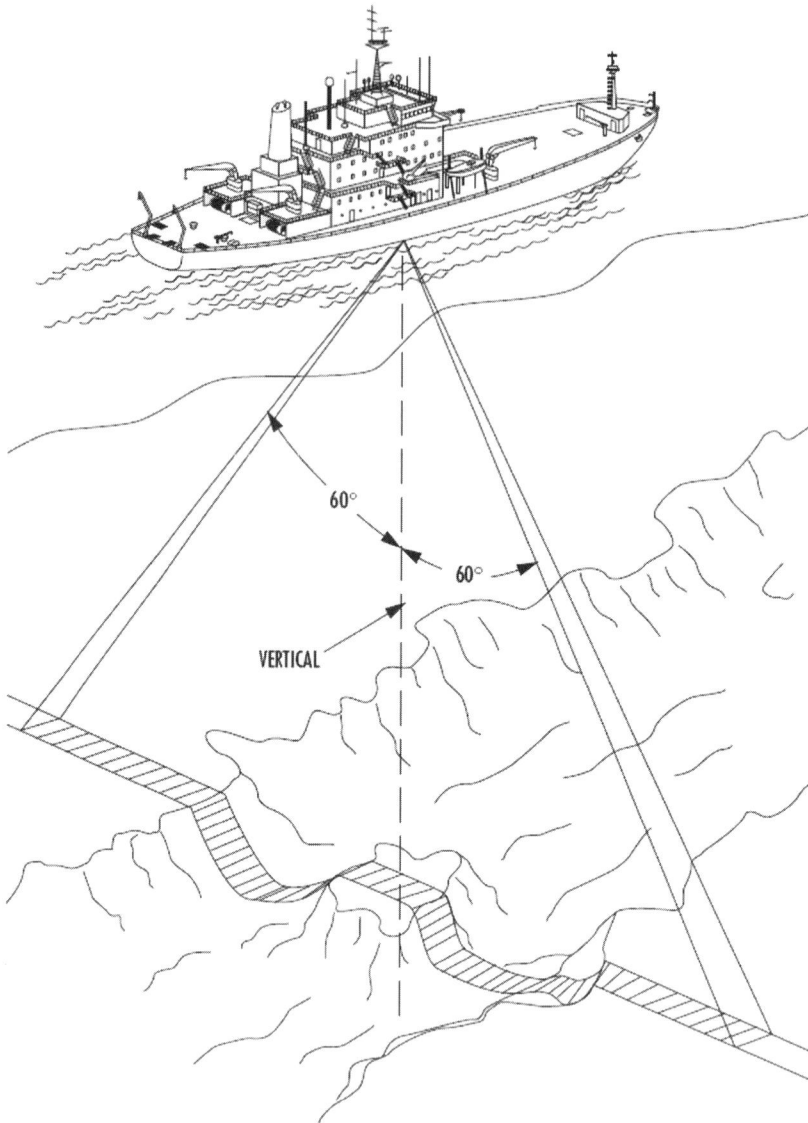

Fig. 1. Swath mapping system.

depths between about 10 and 600 m. The EM1000® is available from the Simrad Company, but similar systems are also constructed by Elac-Nautik, Konigsburg, Reson and GeoAcoustic. These shallow-water systems can provide data at up to sub-metric resolutions in both vertical and horizontal planes.

The data presented in this paper were acquired using low-frequency, deep-water systems that operate at up to full oceanic depths. The system used for the Hawaii and PNG surveys was the SeaBeam 2112® built by L3-Communications

Seabeam Instruments, Inc. SeaBeam 2112® operates at a frequency of 12 kHz, producing a fan of 149 beams with $2° \times 2°$ beam widths, and a total angular swath width of between 90° and 120°; with decreasing beam widths used in increasing water depths. The horizontal resolution of the bathymetric data depends upon the water depth and the ship speed. The vertical accuracy of the depth measurement is reported as 0.5% of water depth. The bathymetric data represent a maximum of 120 data points per sonar ping. The acoustic backscatter data (only available for the Hawaii dataset) contains up to 2000 pixels per ping. The swath bathymetric data for the Sumatra survey were acquired by the Royal Navy hydrographic vessel HMS *Scott*, using a 12 kHz SASS-IV system built in 1964 by SeaBeam Instruments – at the time Harris Anti-submarine Warfare Division of General Instruments Corporation (Fig. 1). This was one of the first systems built, and is still one of the highest resolution systems available. There are 361 beams and a 120° swath width. This gives an internal resolution of a third of a degree across the swath. On the Sumatra survey the theoretical beam width at the seabed (horizontal resolution or 'footprint') was *c.* 25 m directly beneath the ship, increasing to 100 m at the outer beams (50°–60° swath width), assuming a flat seabed at 4500 m water depth. The ideal depth precision (vertical resolution) was *c.* 5 m directly beneath the ship, assuming no position or attitude errors and using Rayleigh's criterion; smaller features were sometimes identifiable due to their coherence over several pings. Minor roll artefacts and noise probably reduced the vertical precision to 10–15 m for outer beams.

Another application of sound transmission in water is based on the variation in seabed material that causes different absorbing and reflecting properties of the incident acoustic pulse. The technology used to map these variations is called sidescan sonar. Hard material such as rock is very efficient at reflecting acoustic pulses, whereas softer material such as clay and silt absorb sound and are weak reflectors. Thus, strong reflectors make strong echoes and weak reflectors weak echoes. Knowing these characteristics allows identification of the general composition of the sea floor. Sidescan sonar employs much the same type of technology as multibeam depth sounding. Sound pulses are transmitted and seabed returns recorded. However, it is the amplitude of the returned sound pulse that is utilized to produce seabed 'reflectivity' maps. The seabed composition determines the amplitude of the returned pulse, with harder material producing higher amplitude returns and vice versa. Thus, rock on the seabed gives a strong high-amplitude return, with the return from sediment being much weaker. As with bathymetric swath mapping, sidescan sonar provides a mosaic of seabed reflectivity that can be used to map seabed composition. In association, multibeam bathymetry and backscatter intensity data provide a powerful tool for mapping the seabed.

Bathymetric DEMs: history of development

Digital elevation models of bathymetric data were first produced during the 1980s when multibeam mapping systems, acquiring data at regularly spaced intervals, first became generally available for non-military use. Up until this time, water depths were measured by single-beam echo sounders that acquired data at points along a single line below the course of a ship. In deep waters, data coverage was sparse and data points often irregularly spaced. In shallow waters, where data were often acquired for marine safety and for the making of bathymetry charts, data coverage was more comprehensive. Thus, the accuracy of bathymetric maps tended to decrease with increasing water depth. Traditionally, bathymetric maps produced from the single-beam data were contoured by hand with contouring regarded as an 'art', although some regarded the contouring as more of a 'prejudice'.

The development of multibeam bathymetry systems led to a revolution in marine mapping. Rather than one acoustic signal, a large number were emitted in a fan shape from a series of transmitters mounted on a ship's hull, attached to a ship's side by arms or booms or on a towed instrument termed a 'fish'. Acquisition of depth data along overlapping paths (swaths) provides complete depth coverage of the seabed surveyed. Swath width is dependent upon the system used, as well as the depth of water, but is generally 4–5 times water depth. In deep waters of 5000 m, data can be acquired in swaths 20–30 km wide. Ship speeds can be as high as 14 knots without loss of accuracy and depending on depth of water and ping rate. Thus, using swath systems, mapping of large areas of the seabed can take place rapidly and with accuracy far better than that using single-beam echo sounders. With precise differential positioning, data correction related to ship movement and, in shallow waters, tidal variation, bathymetry acquired can be rectified to absolute datums. To address the effects of the physical properties of the water column on the sound signals (ray bending), velocity profiles are usually acquired from expendable bathythermographs.

Digital elevation models of relatively small seabed areas are commonly compiled from specific thematic surveys, producing bathymetric mosaics of the seabed. On the global scale, they are also produced for the oceans (Smith & Sandwell 1997). With the declassification of military satellite gravity data in 1995, DEMs of the world's oceans were constructed. The satellite gravity data reflects the topographical variations in the ocean floor and, validated by ship-based echo soundings, provides seabed morphology and bathymetric maps with horizontal accuracies of 1–12 km and vertical accuracies of tens of metres (Smith & Sandwell 1997).

Thus, within decades, seabed mapping has been revolutionized. Based on the application of technology originally developed for military use, together with major advances in data storage and manipulation, the seabed can now be mapped with an accuracy and coverage almost comparable to that on land. Although a primary advance is based on multibeam technology, the development of this technology has been dependent upon an improved navigational accuracy provided by satellite systems and improved computing power. Initially, in the 1960s, TRANSIT satellite navigation was developed with a horizontal accuracy of 100–220 m. The Global Positioning System (GPS) developed in the 1980s now provides a horizontal accuracy of a moving vessel that is measured in metres. The advances in multibeam technology have also been dependent upon the improved computer processing speed and manipulation of the enormous volumes (Gigabytes) of data acquired during multibeam surveys. Without these advances in navigation and computing the full potential of multibeam bathymetric systems could not have been adequately fulfilled.

Multibeam bathymetry: impact and application

In this paper DEMs based on multibeam and sonar data are presented from three areas: off PNG, Hawaii and Sumatra. In Hawaii the data were initially acquired as part of the USA's programme to delimit its Exclusive Economic Zone. Subsequently, its main use has been in researching the hazard from volcanic lateral collapse. The use of sidescan sonar was instrumental in the early identification of this hazard. In PNG and Sumatra swath bathymetric data were specifically acquired to investigate geohazards from earthquakes and tsunamis. The generic objective of the research associated with these investigations has been to determine the hazard of tsunami from submarine landslides or volcanic slope/flank failure. Up until 1998 it was generally assumed that the main source of hazardous tsunami was from earthquakes, with secondary sources including volcanoes and, rarely, bolide impacts. Although submarine landslides were known to have caused some prehistoric events (such as the Storegga submarine landslide, 8500 BP) and small recent tsunami (such the Grand Banks event of 1929 and Seward, Alaska in 1964), these were interpreted as representative of a hazard where the loss of life would be limited. It was in July 1998 that a devastating tsunami on the north coast of Papua New Guinea (PNG) killed over 2200 people (Kawata et al. 1999). The earthquake magnitude of around 7.1 was too small and too early in relation to the tsunami wave inundation to explain the tsunami run-ups measured at up to 15 m. After a focused marine survey programme,

a submarine landslide was identified as the cause of the tsunami (Tappin et al. 1999, 2001, 2002, 2003, 2008). The event changed perceptions as to the threat from tsunami sources from submarine landslides.

In 1998, because the tsunami hazard from submarine landslides was not appreciated, there had been few attempts to model these as a tsunami source. Modelling was based on theoretical considerations (e.g. Jiang & LeBlond 1992, 1994), although some authors used basic landslide models based on single-beam bathymetry; for example the 1975 Kitimat submarine landslide (Murty 1979) and the Storegga event (Harbitz 1992). For the Hawaii volcanic flank collapses some modelling was carried out (Aida 1975; Smith & Shepherd 1996) motivated by the work of Moore and others on the Hawaiian landslides. Modelling of subaerial landslides based on the 1958 Lituya Bay landslide (e.g. Wiegel 1964) had also been undertaken. Other tsunamis that may have involved submarine landslides such as Unimak, Alaska, 1946 and Sanriku, 1896, were considered primarily in the context of their earthquake source (e.g. Abe 1979; Johnson & Satake 1997). In 1998 the landslide constitutive equations used in modelling were largely untested by laboratory experiments or case studies. Submarine mass failure (SMF) models were not based on geological data, but on idealized SMF morphologies. There was no established method of merging geological data with SMF models. In total, there was little appreciation of the complexity of modelling tsunamis generated by different SMF mechanisms. The PNG tsunami was a significant event because of the large loss of life, and also because it gave the impetus to develop improved mathematical models of submarine seabed failure and volcanic flank collapse (e.g. McMurtry et al. 2004b). With multibeam data available, for the first time it was possible to use these data as a basis for the models. For both submarine landslides and volcanic flank collapse, the essential physics of tsunami generation are similar.

When the Indian Ocean Tsunami struck on 26 December 2005, one of the first marine datasets acquired was multibeam bathymetry offshore of northern Sumatra (Henstock et al. 2006). As with PNG, there was little detailed bathymetric data available for the offshore area of the earthquake rupture zone. The high-resolution multibeam data acquired allowed a 'first look' at the seabed morphology, from which the various styles of submarine mass wasting, thrusting and small-scale tectonics could be interpreted (Tappin et al. 2007).

DEMs based on multibeam swath bathymetry presented here have been used to determine whether submarine or volcanic landslides are present in the areas surveyed and, if so, their particular morphology. The term submarine landslide encompasses numerous mechanisms of seabed failure that may be considered by their end members: landslides that are fragmental sediment failures in non-cohesive sediments that have long runouts, and slumps that are cohesive, rotational failures with restricted horizontal movement. In the instance of offshore Sumatra, the data, as well as being used to map submarine landslides, have also been used to identify vertical seabed movement associated with the great earthquake of 26 December 2004. The DEMs from PNG and Hawaii are based on deep-ocean multibeam systems that utilize low, 12 kHz, frequencies that allow mapping at full oceanic depths of up to $c.$ 10 000 m. The main drawback of these multibeams, compared with higher frequency, shallower water systems, is that they have lower vertical resolutions that are measured in tens of metres. However, the deep-water multibeam deployed during the Indian Ocean survey is a military system that, although at a frequency of 12 kHz, has a better vertical resolution of than counterpart civilian systems. In the instance of PNG, we show the improvement in processing and imaging multibeam data that has taken place since 1999, comparing DEMs of contoured data with 3D models of seabed relief developed in the 3D imaging and interactive software *Fledermaus*.

The Papua New Guinea tsunami – 1998

The offshore survey programme in PNG was one of the first to be carried out specifically to identify the cause of a tsunami that had taken place only recently. A major catalyst for the survey was the great loss of life together with the uncertainty over the tsunami source. The comprehensive geophysical dataset acquired off of northern PNG includes over 19 000 km^2 of multibeam bathymetry, 4.2 kHz high-resolution sub-bottom seismic lines, and both single and multichannel seismic data. In the region of the source area of the tsunami, there are also four 7 m-long sediment piston cores, together with shallow (30 cm) sediment push cores, rock samples and marine organisms, together with still and video photography of the seabed acquired by the tethered remotely operated vehicle (ROV) and manned submersible (Tappin *et al.* 1999, 2001, 2002, 2003, 2008; Sweet & Silver 2003). The marine dataset contributes to the understanding of the tsunami source because from it we identify: (i) the background tectonics and sedimentation regime of the area; (ii) the slump and its architecture; and (iii) the relative timing of slump failure. The basis of the interpretation is the DEM prepared from the multibeam bathymetry. Initially, the bathymetric data were imaged as a 2D contoured map (Fig. 2a). Subsequently, the data were imported into the more sophisticated interactive imaging software *Fledermaus*, thereby significantly improving visualization and the interpretation of the multibeam data.

Contoured bathymetry

From the contoured multibeam data the main features of seabed morphology may be identified (Fig. 2a). The data reveal a transpressive convergent margin, with the North Bismarck Sea and Caroline plates colliding with northern PNG along the New Guinea Trench. In the east, collision of the North Bismarck Sea Plate, a significant seabed high, is causing tectonic erosion along the inner trench slope. The data reveal a steep and narrow inner trench slope with back-tilted fault blocks formed by collapse along the trench. The morphology is rugged and the seabed deeply dissected by canyons. In the west, the inner trench slope is wider, because of the presence of a series of lower slope basins (absent in the east because of tectonic erosion). As in the east, on the upper slope, there are back-tilted fault blocks. Minor sediment depocentres are located off major rivers such as the Bliri and the Pual, but the rugged topography indicates that there is little sediment entering the area. Only at the mouths of rivers, which are few, is there any evidence for sediment build-ups. There are both extensional and compressional features including slumping, faulting and uplifted blocks.

Based on the multibeam data, it is apparent that there are two morphotectonic regions. At their boundary, and located 20 km offshore of Sissano Lagoon (the area devastated by the tsunami), is the feature termed the 'Amphitheatre' (Figs 2 & 3). The arcuate shape of the Amphitheatre indicates formation by submarine slope failure, probably along a control fault located along the base of the headscarp (Fig. 3). In the centre of the Amphitheatre lies a slump. The main features of the Amphitheatre can be identified on the contoured bathymetry maps. The slump is identified by the steep headscarp slope and a basal mound at the slump toe. The relatively smooth surface morphology of the basal mound contrasts with the seabed morphology in the western part of the Amphitheatre, which is irregular and deeply incised by seabed gullying. The contrasting morphology within the Amphitheatre indicates that the slump is restricted to a discrete area in the centre, indicating a rotational cohesive failure rather than a translational landslide. Other details of slump morphology were identified from seismic data and seabed photography. From the dataset, the slump dimensions were set at 5 km long in a north–south direction and 7 km wide.

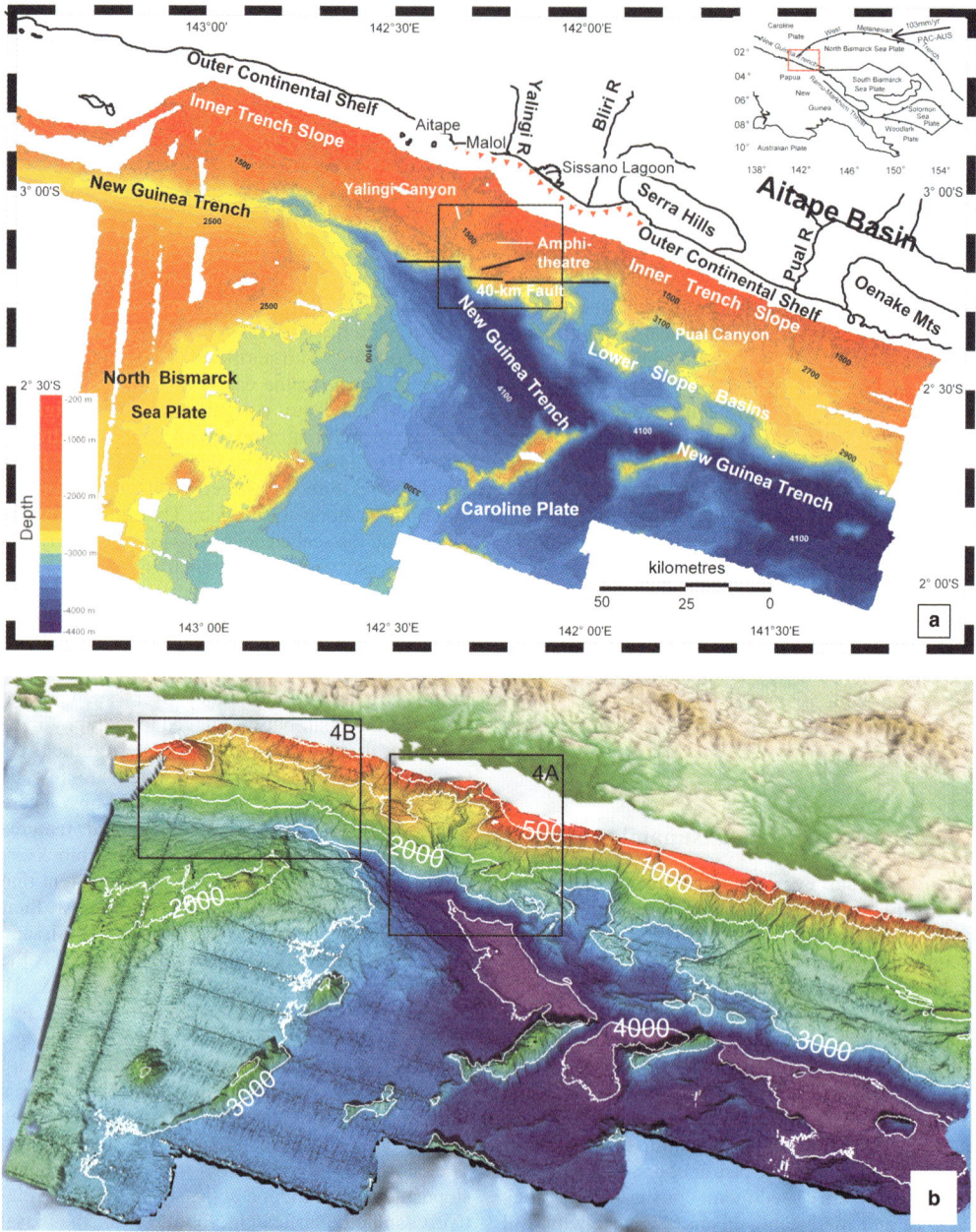

Fig. 2. PNG regional bathymetry viewed from the north. (**a**) Contoured bathymetry and main morphological elements offshore of northern Papua New Guinea, together with coastal locations and features. Red triangles show the main area devastated by the 17 July 1998 tsunami. The box is the area of Amphitheatre shown in Figure 3a. Inset map: location of the mapped area (red box) with the main tectonic elements and plate motions indicated. (**b**) Three-dimensional DEM of sea-floor relief with bathymetric contours (vertical exaggeration ×4). The boxes show locations in Figure 4.

With regard to the faults in the vicinity of the Amphitheatre (Fig. 3a), these are dominantly dip-slip; there is no evidence to support the presence of major thrusts. The absence of evidence for thrust faulting supports the interpretation of the earthquake as a shallow-dipping, blind thrust with no surface expression. The most significant fault in the region of the Amphitheatre is the 40 km fault,

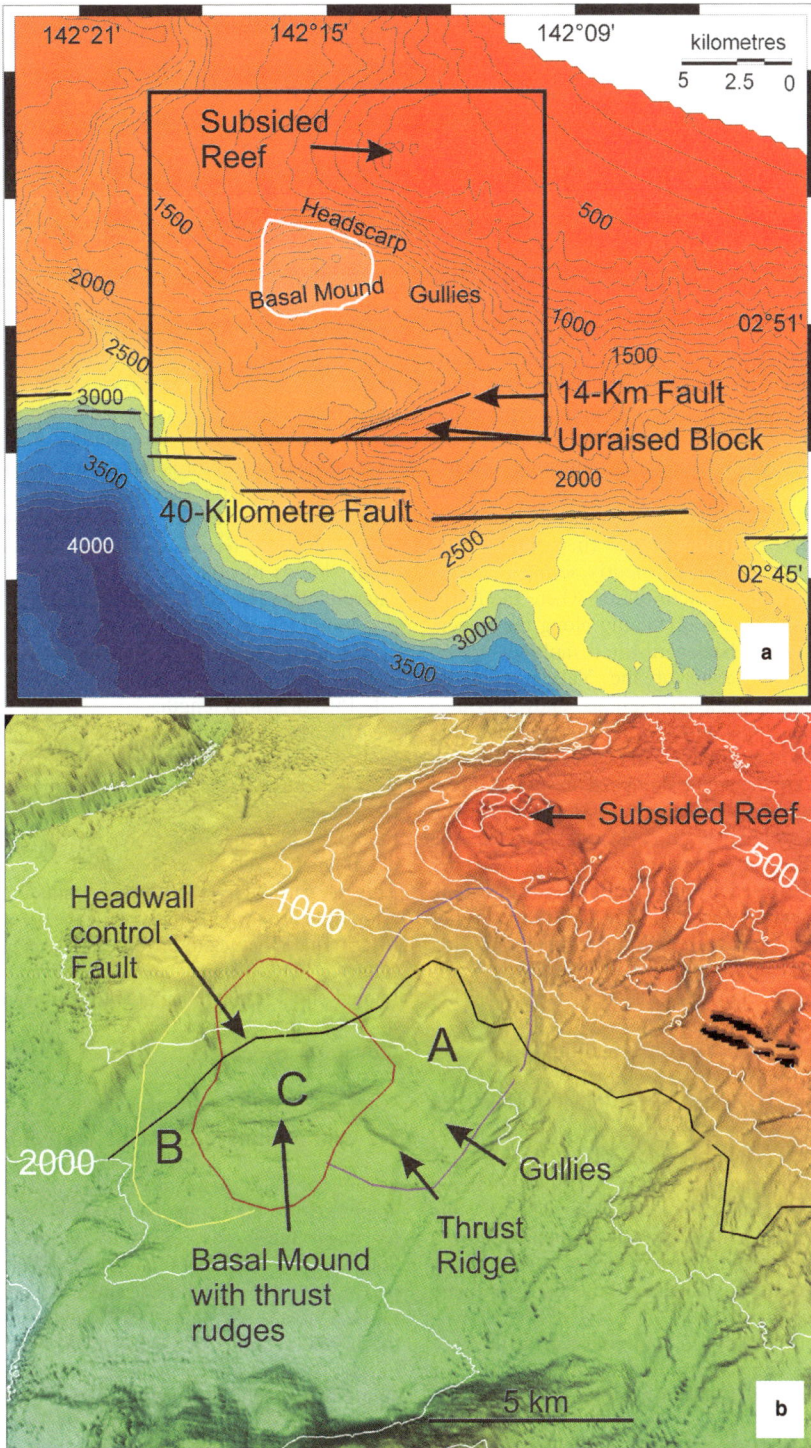

Fig. 3. The PNG Amphitheatre region viewed from the north. (**a**) Contoured bathymetry (location in Fig. 2a) with main morphological features labelled. Solid black lines are faults. The white hachured area defines the main slump area of 17 July 1998. The box is the area of Figure 2b. (**b**) Three-dimensional DEM of sea-floor relief with bathymetric contours and main seabed features (vertical exaggeration ×4).

located north of the Amphitheatre. The movement along this fault is dip-slip, with a downthrow to the north. The 14 km fault lying within the Amphitheatre is mainly a normal fault, although most recent fault movement is suspected to be in a reverse direction.

DEMs of seabed relief

The DEMs of seabed relief, by comparison with the contoured data, provide a much clearer image of the mapped area (Figs 2b, 3b & 4). There is better definition of the backward-rotated fault blocks (Fig. 4b). Strike–slip faulting is clearly seen to intersect the canyons dissecting the lower part of the inner trench wall. Below the slope-parallel depressions formed by the backward-rotated faulting the canyons are offset to the east, indicating sinistral movement, as would be expected to result from the oblique convergence taking place between the Caroline and Australian plates (inset in Fig. 2). The strike–slip faulting is clearly seen to be confined mainly to the eastern region where the North Bismarck Sea Plate is impacting along the trench. The effect of the backward-tilted faulting is seen in the sinuous nature of the Yalingi Canyon, which is turned back on itself to landwards at a number of locations (Fig. 4a). The heavily dissected nature of the inner trench wall is exceptionally well imaged, as are the small slumps. The subsided reef off of Sissano Lagoon lying at 300 m water depth is clearly visible (Figs 3b & 4a) and indicates the extent of the subsidence taking along the inner trench wall caused by the tectonic erosion along the convergent margin boundary. The seabed relief DEM confirms that there is little sediment entering the offshore region from the land. There are numerous submarine canyons, but they are delivering little sediment to the area.

On the oceanic plate step faulting on the SW margin of the North Bismarck Sea Plate as it descends into the Trench is clearly imaged (Fig. 2b). Along the base of the inner trench wall there is no evidence for young accretionary thrust folds formed in sediment offscraped from the subducted plate (cf. the Indian Ocean images later in this paper) confirming that, not only is there a limited supply of sediment entering the area from land, but also very little is entering the subduction zone from the oceanic plate.

Within the area of the Amphitheatre, the DEM of seabed relief (Fig. 3b) provides an improved level of detail that allows the identification of three phases of slumping (labelled A, B and C in Fig. 3b). The seabed morphology also allows the relative age of these slumps to be identified. The heavily gullied seabed in the west, in contrast to the seabed in the centre that is not gullied, indicates that this area

has not failed as recently as the central area. Here slump A is identified by its heavily gullied surface, with arcuate thrust ridges on its surface (Fig. 3b). Its headscarp can be traced to the foot of the subsided reef. The most easterly slump B (identified by Sweet & Silver 2003) is probably younger than the slump in the west, as its surface is not gullied. The central slump C is the youngest as it cuts the two slumps on its margins. It is this slump that is interpreted as failing on 17 July 1998, thereby causing the tsunami of that date. On the surface of slump C, on the elevated mounded region below the headscarp, there are curvilinear, but generally east–west-trending, ridges and furrows that are convex towards the north, and clearly terminate at the eastern and western boundaries of the slump toe. These features are the surface expression of small thrust faults (or pressure ridges) that are imaged on the seismic data and which formed during downslope slump movement. This youngest slump terminates the thrust ridges of the older slumps to both the east and west. The new interpretation of the slump reveals a width of c. 5 km, a length of about 5 km and a thickness of c. 750 m. The slump volume is estimated to be around 6.4 km^3. Slump thickness is normally about 10–15% of slump length.

The overall improvement in the quality of the seabed relief DEM compared with the contoured data is manifest. The main morphological features of the mapped area can be interpreted from the contoured bathymetric maps, but there is more fine detail visible on the seabed surface relief DEM. This is particularly the case in the area of the Amphitheatre, where the data have been used to discriminate between different phases of seabed failure. There is a great deal of other data (seismic, seabed photographs) available to confirm the above interpretations, but it is the DEM upon which this is primarily based.

Hawaiian giant submarine landslides

Large-scale, gravitational collapse of the volcanic slopes of the Hawaiian Islands has been inferred for many years (see review by Moore & Clague 2002). The evidence of collapse is mainly based on the presence on land of seaward-facing normal faults (Stearns & Clark 1930) and, offshore of the northern coasts of Oahu and Molokai islands, on the irregular submarine morphology mapped by single-beam echo sounders (Moore 1964). However, it was not until the 1980s, when the United States of America mapped its Exclusive Economic Zone, that the size and extent of these collapses, termed giant submarine landslides (GSLs), was fully appreciated (Fig. 5). Between 1986 and 1991,

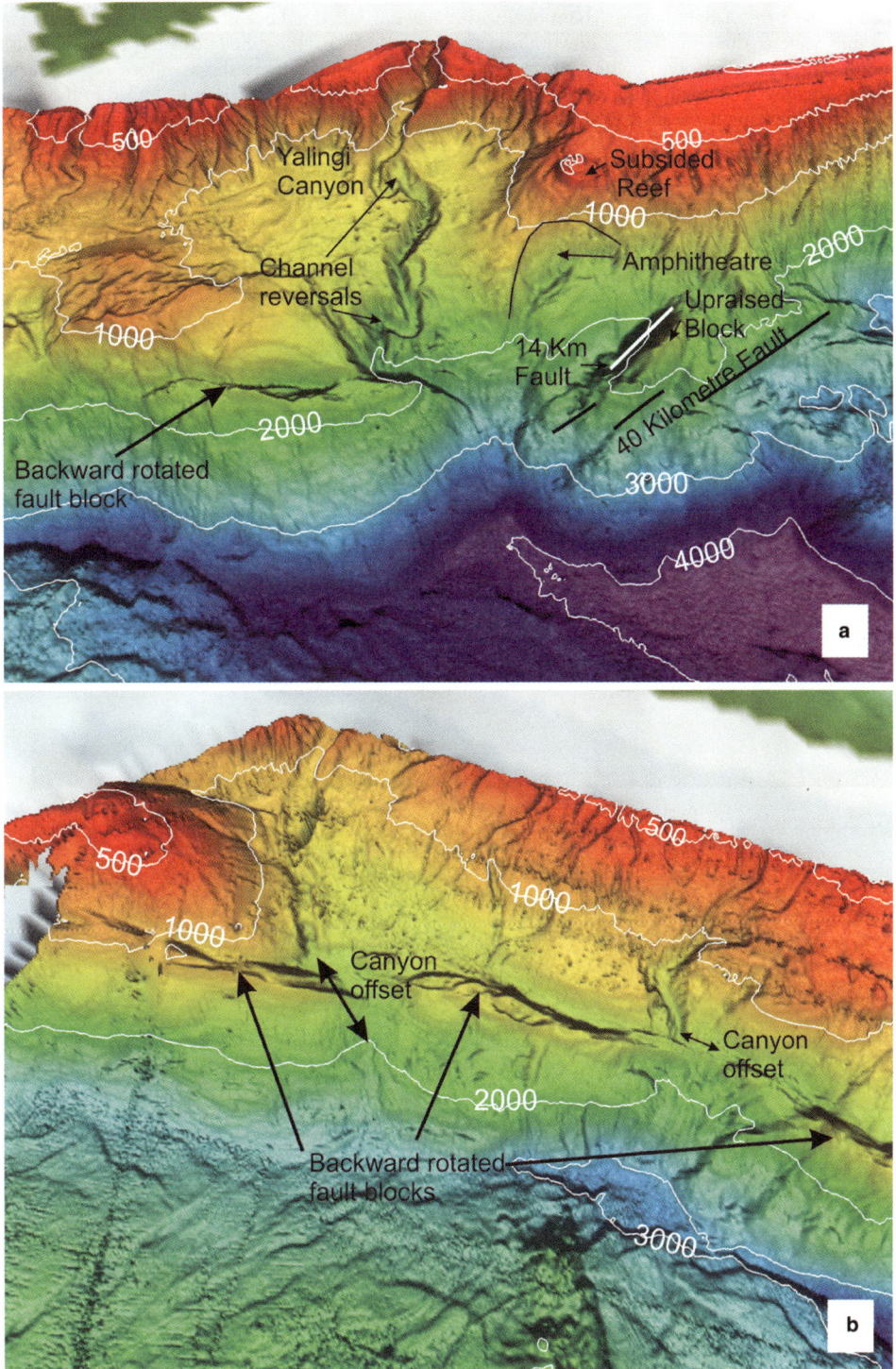

Fig. 4. Details of the PNG regional morphology (locations shown in Fig. 2b). (**a**) Yalingi Canyon and Amphitheatre region. (**b**) Eastern region showing backward-rotated fault blocks and submarine canyon offsets (vertical exaggeration ×4).

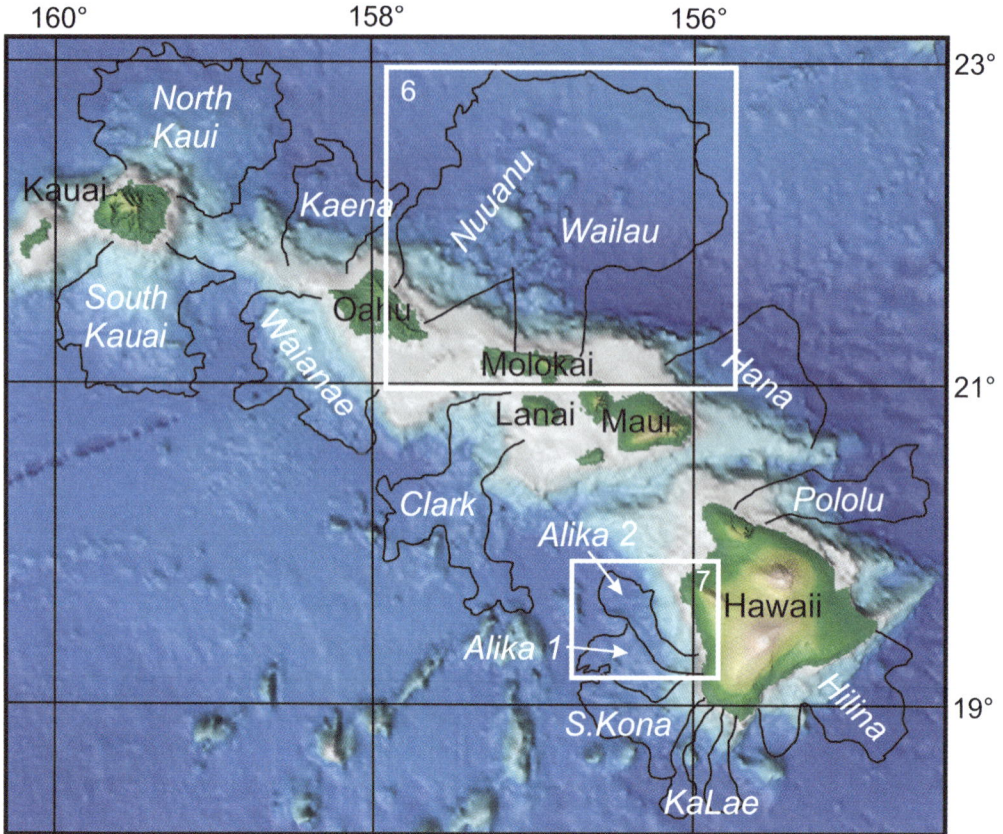

Fig. 5. Hawaiian seabed morphology, and locations and extents of the giant submarine landslides as mapped by Moore *et al.* (1989). The boxes show areas of Figures 6 and 7.

using the 6.6 kHz GLORIA sidescan sonar, 68 major landslides over 20 km in length were identified offshore of the Hawaiian Islands (Moore *et al.* 1989, 1994). Some of the landslides are over 200 km long, with volumes of up to 5000 km^3. This was the first time that such large-scale features had been identified, and subsequently they proved to be the largest landslides on Earth.

The GLORIA backscatter data, together with single-beam bathymetry, reveal two main landslide morphologies, debris avalanches and slumps, with numerous variations in-between. The majority of the failures are debris avalanches, and these lie on slopes of less than 3° and are formed of fragmented volcanic rock. The debris avalanches are hundreds of metres in thickness, with well-defined amphitheatres at their head. Their hummocky surface topography is comparable to similar subaerial deposits, such as those from the Mount St Helens collapse of 1980 (Lipman *et al.* 1988). Their morphology, together with their large lateral extent and the fact that they travelled long distances across the Hawaiian Deep

(Fig. 6a) upslope onto the Hawaiian Ridge, in places overtopping vertical elevations of hundreds of metres, indicates that the GSL travelled at high velocities (100–200 m s^{-1}: e.g. McMurtry *et al.* 2004*b*). The debris avalanches are thus probably catastrophic because they take place as single events. The slumps, in contrast to the debris avalanches, are located on slopes with gradients of more than 3° and are up to 10 km in thickness. They failed slowly by creep, resulting in little internal deformation. Their heads extend back to the volcanic rift zone, whereas their bases extend down to the root of the volcanic pile.

The discovery of the GSLs led to a major debate on whether these massive failures could create hazardous tsunamis (Moore & Moore 1984, 1988; Rubin *et al.* 2000; McMurtry *et al.* 2004*a*, *b*). Initially, the evidence rested on marine calcareous gravels at elevations of up to 300 m above sea level found on the islands of Hawaii, Lanai and Molokai. When these sediments were first discovered in the 1940s they were interpreted as

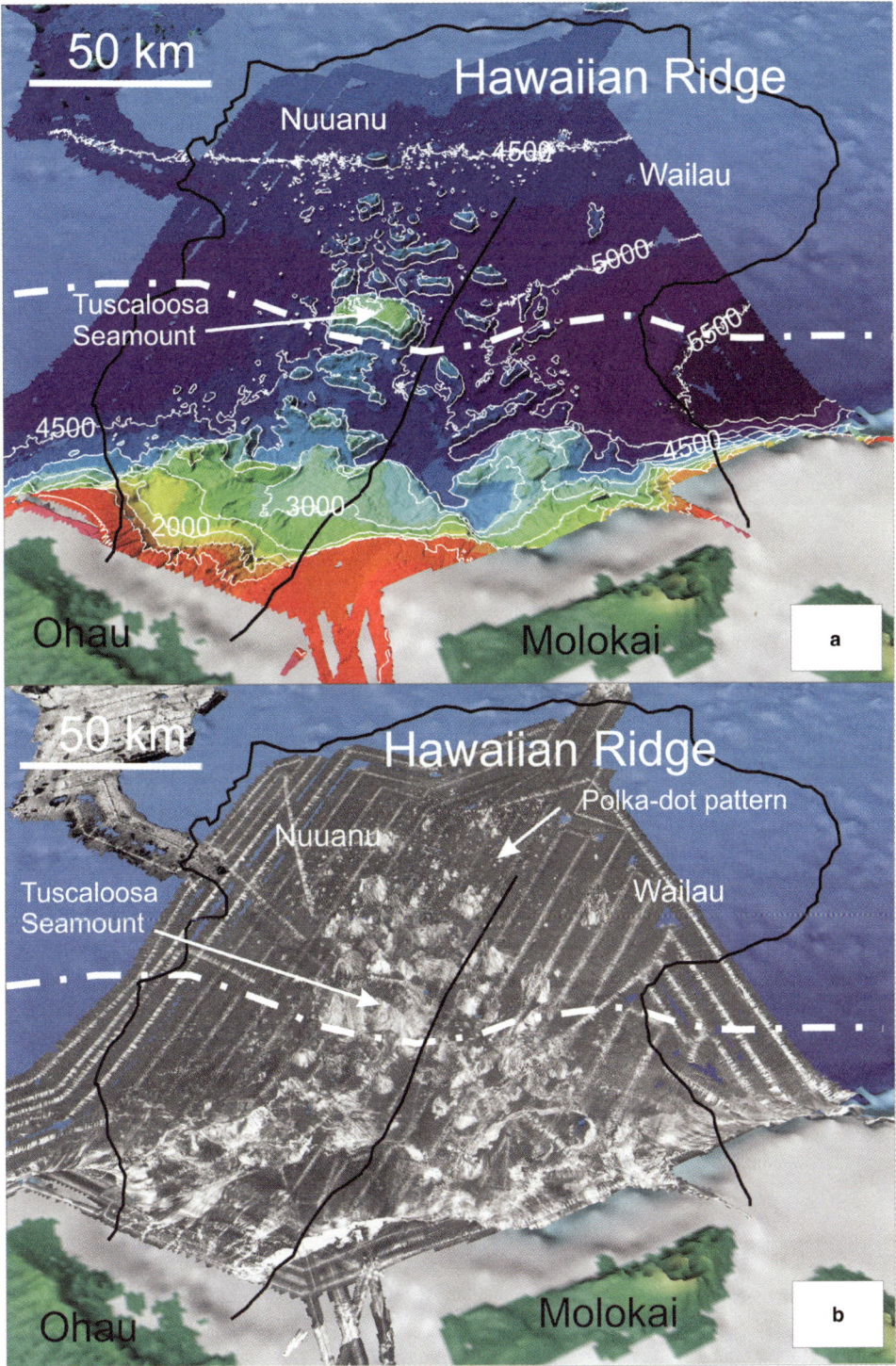

Fig. 6. Hawaiian giant submarine landslides – Nuuanu and Wailau. (**a**) Three-dimensional DEM of sea-floor relief with bathymetric contours. Black lines define the boundaries of the GSL. The dashed white line is the axis of the Hawaiian Deep. (**b**) Backscatter seabed image (vertical exaggeration ×4).

deposited during sea-level highstands (Stearns & MacDonald 1946; Stearns 1978). When the GSLs were discovered it was proposed that the sediments were, in fact, laid down by the tsunami caused by these massive GSL events (Moore & Moore 1984, 1988). The arguments have swung back and forth, and after a more recent reinterpretation of the conglomerates as highstand deposits (Rubin et al. 2000), modelling of GSL failure indicates local run-up heights of hundreds of metres (McMurtry et al. 2004b). Recent analysis of calcareous conglomerates on Hawaii indicates that these were most probably laid down by a tsunami with a vertical run-up of over 300 m above sea level at the time of deposition (McMurtry et al. 2004a).

The GLORIA backscatter images provide the morphology of the seabed features but not water depths, which were acquired using single-beam echo-sounders. In the late 1990s a new series of marine surveys acquired multibeam data over many of the GSLs (Takahashi et al. 2002; Smith et al. 2002). This new dataset includes 12 kHz Seabeam multibeam and backscatter data (acquisition and processing parameters to be found in Smith et al. 2002). In association with seismic data, that penetrates the seabed, the 3D architectures of the GSLs have now been mapped in greater detail than previously (Fig. 5). The focus in this paper is on two areas offshore of the Hawaiian Islands where there are several large GSLs: the area north of Oahu and Molokai; and offshore of the west coast of Hawaii. In the former area are located two of the largest landslides, Nuuanu and Wailau (Fig. 6). Off the west coast of Hawaii, there is the Alika landslide (Fig. 7), the youngest of which, Alika 2 at 120 000 years BP, created the tsunami that laid down the controversial sediments hundreds of metres above sea level (at the time of deposition) discussed earlier.

The Wailau and Nuuanu debris avalanches

The Wailau and Nuuanu landslides are debris avalanches resulting from the collapse of the northern coasts of Oahu and Molokai (Figs 5 & 6) (see Moore & Clague 2002). Their limits, as well as their common boundary, are imperfectly defined. Two main regions have been identified: an inboard area where there are large slipped blocks, outboard of which lies an irregular seabed identified on the backscatter images by a 'polka dot' pattern that reflects a hummocky terrain (also seen on the bathymetry) formed by relatively smaller blocks set in unconsolidated sediment (Fig. 6). Outside of these regions the exact margins of the landslides are difficult to define because the fine-grained sediment apron of the GSL is hard to distinguish from the background hemipelagic sediment. The Nuuanu

debris avalanche is the largest GSL in Hawaii. It is 230 km long and covers an area of 23 000 km^2, with a volume estimated to be between 3000 and 5000 km^3 (Moore et al. 1989; Satake et al. 2002). It crosses the axis of the Hawaiian Deep and moves upslope a distance of 140 km, rising to a vertical elevation of more than 300 m (Fig. 6a). The landslide blocks within the debris avalanche are up to 700 km^3. The Tuscaloosa Seamount rises to over 2 km above the surrounding seabed. Block size decreases outwards from the landslide head. In total the blocks comprise a volume of 1400 km^3, almost 50% of the landslide volume. The Wailau debris avalanche is about the same length as Nuuanu, but is smaller in volume at 1400 km^3.

Dating of igneous rocks forming the blocks and turbidites in piston cores shows the Wailau debris avalanche to be about 1.5 Ma, with the Nuuanu debris avalanche about 1 Ma older (Kanamatsu et al. 2002; Shinozaki et al. 2002). Identification of the boundary between the two events has been based on the relationship of the larger blocks to their proposed source regions. This has led to some uncertainty along their common boundary. For example, there are three large blocks along the boundary of the Wailau debris avalanche but with an orientation (with their long axis roughly trending east–west) that suggests they are part of the Nuuanu event (Fig. 6).

The architecture and volume of the Nuuanu and Wailau debris avalanches form the basis of their modelling as tsunami sources (Satake et al. 2002). Assuming that failure takes place during one event, an assumed failure velocity of 50 m s^{-1} and the landslide volumes stated above, the tsunami run-ups on the nearest Hawaiian islands are up to 100 m for the Nuuanu event and up to 75 m for Wailau. In the far-field the highest run-ups of tens of metres are on the southern California coast of the western USA and the Aleutian islands. As noted above, before the 1998 PNG tsunami occurred landslides were not considered a significant tsunami hazard. There is now a general agreement that in the near-field this understanding is wrong and that local run-ups from landslides, both subaerial and submarine, can present a real threat. It is in the far-field, away from the local source area on the tsunami, that the tsunami hazard is uncertain. Previously to recent research, in the far-field, landslides were usually assumed not to be a significant tsunami source. This is despite the evidence from smaller volume (a few km^3) historical volcano collapses, such as Oshima-Oshima in 1741 (Satake & Kato 2001; Satake 2007) and Ritter Island in 1888 (Johnson 1987; Ward & Day 2003), that resulted in tsunamis which created damage up to 1200 km distant (Oshima-Oshima) and more than 600 km distant (Ritter Island) from the tsunami source.

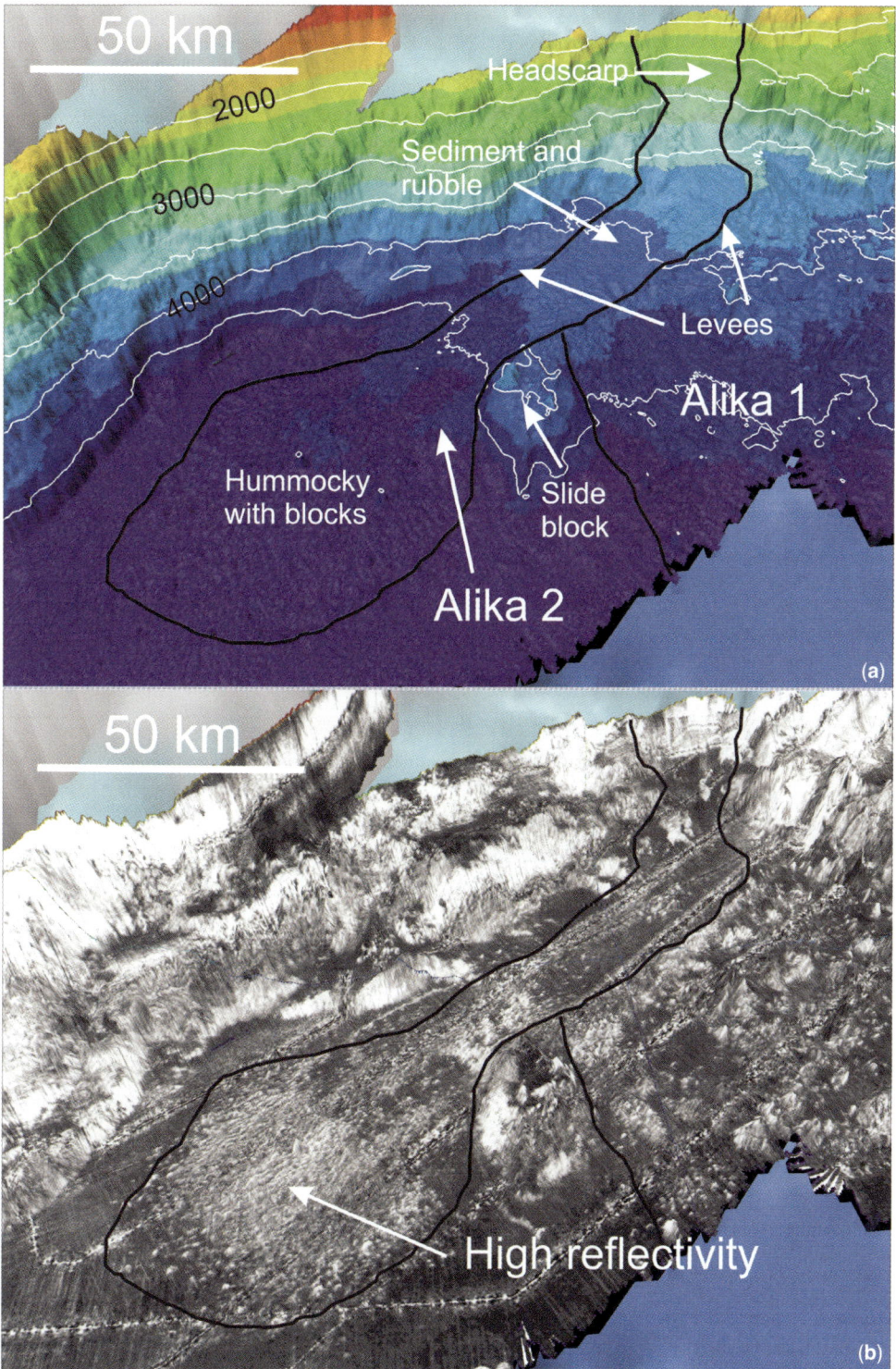

Fig. 7. Hawaiian giant submarine landslides – Alika 1 and 2. (**a**) Three-dimensional DEM of sea-floor relief with bathymetric contours. (**b**) Backscatter seabed image (vertical exaggeration ×4).

The modelled run-ups of Satake *et al.* (2002) resulting from the Hawaii flank failures suggest that the far-field threat is real. It is the enormous size of the Hawaiian events that results in the far-field tsunami. The results of other studies referenced here, together with the modelling of Satake *et al.* (2002), indicates that a re-evaluation of the far-field tsunami threat from Hawaii flank collapse is necessary.

The Alika debris avalanche

To the west of Hawaii there are a number GSLs (Fig. 5), one of the youngest of which is the Alika debris avalanche (Lipman *et al.* 1988) (Fig. 7). The debris avalanche is divided into two events, with Alika 2 being the youngest as it overlies Alika 1. Alika 2 has a total area of 4000 km^2, with a volume of *c.* 200 km^3. All except its upper part has been mapped using 12 kHz multibeam and backscatter intensity data (Fig. 7). Alika 2 is relatively narrow (at 9 km) in its upper part, where the surface topography is hummocky–planar, representing a mixture of sediment and rubble. The margins of the failure are raised levees up to 100 m high. Where a large slide block constrained

the passage of the debris avalanche, an increase in water depth between 4150 and 4400 m marks the boundary between an upper and a lower section. Beyond the slide block, the width of the lower section of the debris avalanche increases and here, at depths of 4700 m, lies the toe of the failure. The surface of the toe is irregular, and comprises large blocks. On the backscatter image the reflectivity is high, indicating an irregular hard surface (Fig. 7b). On the margins of the high-amplitude region there is the 'polka dot' reflectivity observed on the Nuuanu and Wailau GSLs, and indicative of isolated outrunner blocks. Alternatively, the reflectivity may represent larger blocks within a small-block debris avalanche deposit or a debris flow deposit, with a sediment drape on its relatively smooth top surface to account for the lower backscatter. The concordant axes of the two sections are interpreted as indicating one event, with the lower part failing as a fluidized run-out and the upper part a more viscous second phase.

Modelling of a tsunami created by the Alika 2 debris avalanche, based on the seabed mapping data, results in local run-up on nearby Hawaii of over 300 m interpreted height at the time of deposition (McMurtry *et al.* 2004b). These run-ups

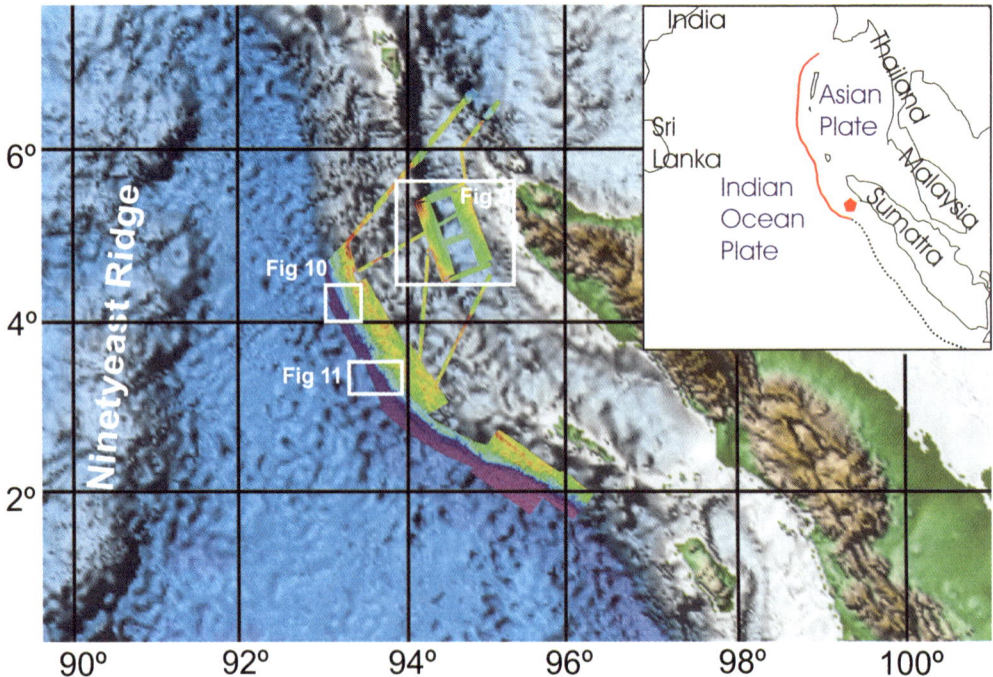

Fig. 8. Map of the marine area off Sumatra and the location of the HMS *Scott* bathymetry data. Boxes are the areas in Figures 9–11. Inset shows the 26 December 2005 rupture (in red) and the earthquake epicentre (red hexagon) (vertical exaggeration ×4).

correspond to the interpreted height of the calcareous gravels on the Kohala Peninsula in northern Hawaii (McMurtry *et al.* 2004*a*).

Indian Ocean: offshore Sumatra

The 26 December 2004 earthquake in the Indian Ocean was the world's largest for over 40 years and created the most devastating tsunami ever recorded, with over 220 000 fatalities. Earthquakes are a commonly cited mechanism for triggering submarine landslides (e.g. Hampton *et al.* 1996; Lee *et al.* 2007); thus, run-ups of over 35 m reported from Banda Aceh in northern Sumatra, close to the tsunami source, might have been enhanced by local submarine landslides (Tappin *et al.* 2007). In the Aceh Basin just offshore of Sumatra (Fig. 8), the existing bathymetry obtained from the General Bathymetric Chart of the Oceans (GEBCO) provided some evidence for seabed sediment failures; with arcuate features identified similar to those found in the Amphitheatre off of PNG described earlier. However, at the time of the tsunami, knowledge of the water depths off Sumatra was generally poor because of the sparse coverage of single-beam echo soundings. The locations of most of these data were uncertain because they were acquired before satellite navigation became available.

Immediately after the tsunami, in January 2005, multibeam bathymetry was acquired offshore of northern Sumatra (Fig. 8) (Henstock *et al.* 2006). This was the first sea-floor survey immediately after a great subduction-zone earthquake, and an ideal opportunity to identify coseismic deformation features in soft sediment. The objective of the survey was to investigate processes associated with major earthquakes at subduction zones using sea-floor morphology; especially to identify vertical seabed movement associated with earthquake rupture and processes of sedimentary mass wasting. The two main areas mapped were the Aceh Forearc Basin, lying offshore of northern Sumatra, and an *c.* 550 km-long section of the central Sunda convergent margin, including the outer-arc high fault system (Fig. 8). It is in the southern part of this region that the December 2004 earthquake epicentre is located.

Aceh Basin

The Aceh Basin is a forearc basin lying 45 km off of northern Sumatra (Figs 8 & 9). Its western boundary is the strike–slip West Andaman Fault (Curray 2005). In the east lies the shelf off of Sumatra. The basin trends NNW–SSE, the basin floor is planar, slightly sloping to the SSW and with water depths of *c.* 2500 m. Seabed gradients on the basin

margins vary between 6° and 12°. The multibeam data prove that the evidence for sediment failures on the eastern margin of the basin identified on the GEBCO bathymetry are artefacts, owing to the sparse coverage of poorly located single-beam echo soundings. There is no evidence for large-scale landslides or slumps that may have contributed to local run-ups during the tsunami. The data do, however, show incised submarine channels on the eastern basin margin, indicating several phases of downcutting. There is no evidence for the channels extending on to the basin floor nor of any significant sediment build-up, such as sediment fans, at their points of entry. There are no channels on the western margin of the basin along the Andaman Fault. Thus, any sediment flow into the basin is not very recent. Sediment transported into the basin through channels is by small-scale sediment flows and any deposition is mainly hemipelagic.

Outer accretionary prism

The multibeam data acquired along the plate margin cover the toe of the accretionary prism up to 75 km inboard of the plate boundary (Fig. 8) (Henstock *et al.* 2006). The lower part of the prism is defined by a rapid change in water depth from 4300–4900 m at the deformation front to *c.* 1500 m on the broad plateau at the top of the slope. The steep lower slope is round 20 km wide, with mean slope gradients in excess of about 8°. There are two morphologies present: those sections with thrust folds (comprising 70% of the margin mapped) and those without. Two main types of mass failure are recognized: blocky debris avalanches and sediment flows, with the majority of failures small-scale.

Blocky debris avalanches. On the toe of the accretionary prism, located on the young thrust folds, there are planar erosional scars that on the bathymetry are usually associated with hummocky seabed topography. The hummocks are interpreted as outrunner blocks derived from the thrust folds. The failure scars are typically ellipsoidal, although some exhibit linear side margins. These features represent sediment failures that are particularly common on the seaward limbs of the thrust folds. At some locations the associated slipped blocks lie outboard of the folds on the ocean basin, at others they are on the surface of upraised sections of the detached young thrust folds or lie between them and the main body of the accretionary prism. A prominent example, and probably the youngest, is found in the north of the mapped region (Fig. 10). This is an 18 km-wide, semi-elliptical slump scar on the outboard limb of a young fold. At its base, on the abyssal plain (the trench has little morphological surface manifestation at this location), the

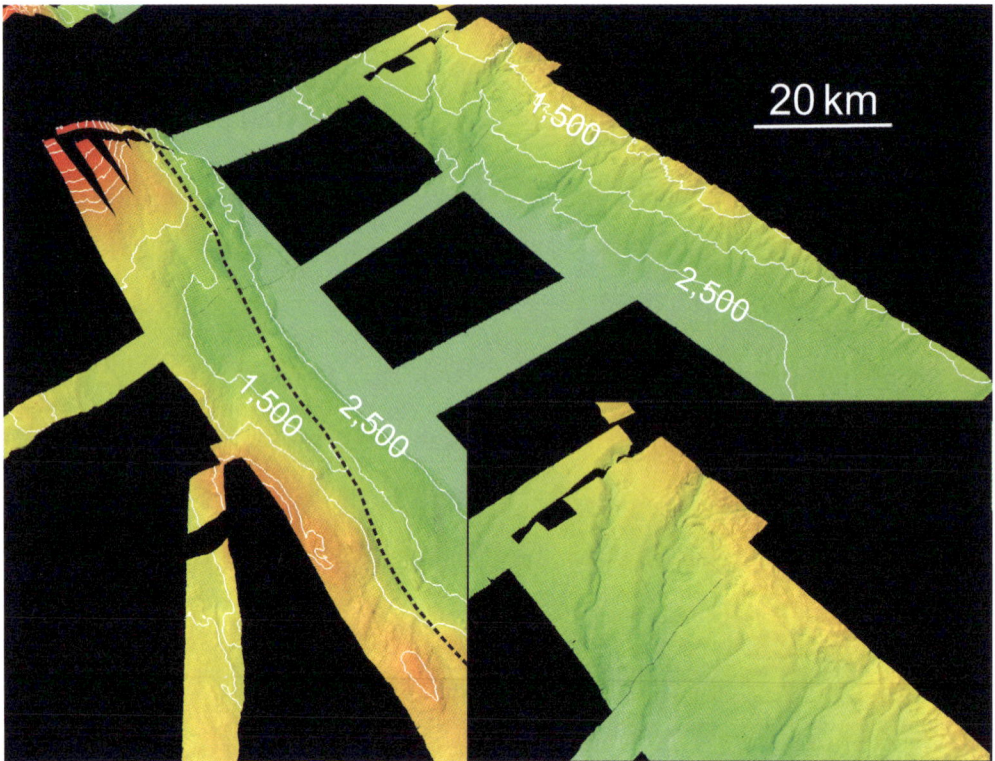

Fig. 9. Morphology and bathymetry of the Aceh Basin. Bathymetric contours in metres. The black dashed line is the West Andaman Fault. Inset shows an enlargement of the channels on the NW margin of the basin (for the location see Fig. 8) (vertical exaggeration ×4).

thrust fold lies at a depth of 4400 m with a crest at 3200 m. The outboard fold limb slopes at an angle of 11°–12°, but at the crest the slope is up to 23°. The failed area corresponds to the steepest gradient and greatest elevation along the fold. The top of the headwall scarp (see Hampton *et al.* 1996 for terminology) lies on a notable, sharply defined, spine on the culmination of the fold ridge. Within the scar there are three areas of mass failure (Fig. 10), with a total thickness of sediment excavation of *c.* 100 m. Measurements of the vertical steps at the boundaries between the three areas of failure indicate that the individual layer thicknesses vary between 20 and 35 m. The outrunner blocks occupy a triangular-shaped area seaward of the fold, with the furthest block outboard forming the apex. The largest outrunner block is about 100 m proud of the seabed and up to 2 × 1 km in length/ width; it lies furthest from the source, with its outboard face lying 10 km from the foot of the thrust fold.

The blocky debris avalanches on the seaward faces of young thrust folds are interpreted to be formed in cohesive, but relatively unconsolidated, sediment. The mechanism of failure is interpreted to be due to a combination of factors; primarily tectonic oversteepening of thrust fold limbs, with the slump margins controlled by movement along small-scale thrusts activated during earthquakes. There is no evidence of fluid expulsion on the DEM, such as might be indicated by pockmarks or mud volcanoes. In addition, no evidence of fluid expulsion was found during ROV surveys in the area (Moran & Tappin 2006). The bedding inherited from the original depositional sediment character controls the thickness of the failed blocks. Failures take place on the steepest slopes, and internal structures seen on the failure scars may not represent individual failures that are widely separated in time. In the instance of the blocky debris avalanche described in detail, although there are three internal scars, these probably failed during one episode of movement. There is a local, small-scale, structural control on the location of the failure. These we interpret to reflect larger, deeper-seated seaward-dipping thrusts structures reflecting landward vergence (Henstock *et al.* 2006).

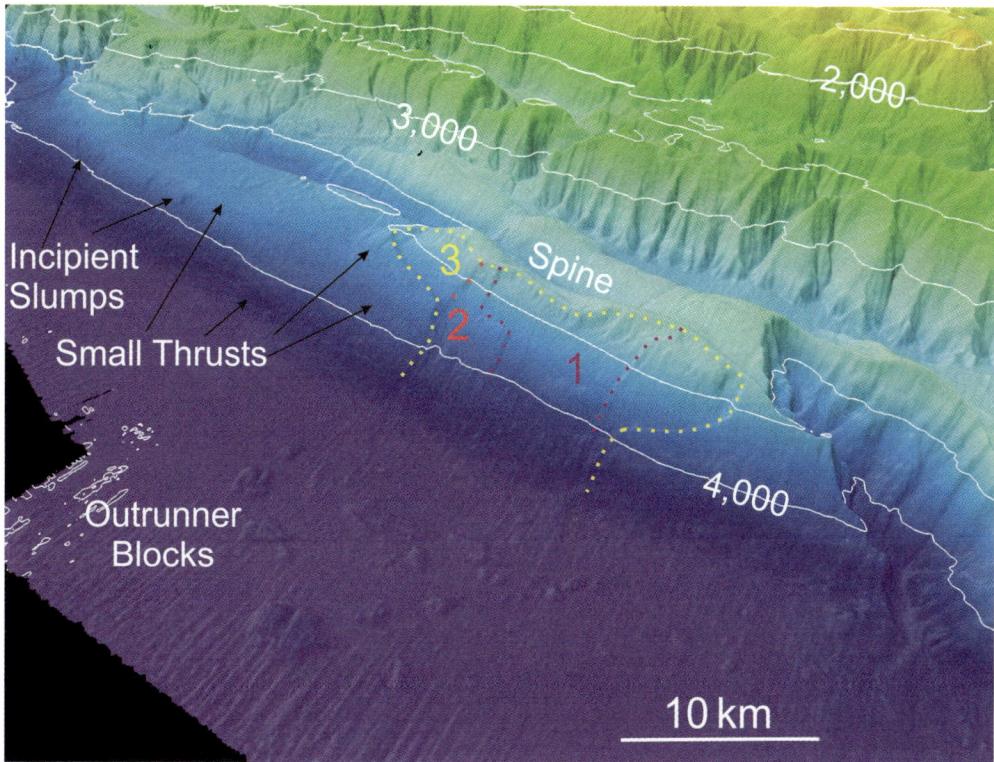

Fig. 10. Morphology and bathymetry of the 'young slump'. Bathymetric contours in metres. Internal dotted lines and numbers within the young slump represent the three internal divisions (for the location see Fig. 8) (vertical exaggeration ×4).

Sediment flows. Along the toe of the accretionary prism, where the young thrust folds are absent, the prism toe rises abruptly from the abyssal plain and seabed gradients approach 30° (Fig. 11). The outboard slopes are heavily incised by numerous gullies. Landwards, the gullies cut through the older thrust folds and lead into arcuate, incised 'cauliform' features (described because of their similarity to a cauliflower) that are similar in form to stream catchments in mountainous regions. Seawards, the gullies lead into channels on the abyssal plain, which are up to 100 m deep. The channel morphology varies; some channels are linear, others are meandering. In some locations at the mouth of the gullies there are small (10–20 m high) sediment blocks. Meandering channels are commonly seen to have a number of episodes of activity. On the abyssal plain there are sediment waves and, at one location, a sediment fan has been formed.

The deeply dissected, steeply sloping, gullied morphology, formed where the young thrust folds are absent, is interpreted to be a result of incremental sediment failure, mainly through headwall erosion.

There is initial failure of larger blocks of sediment in the source regions that break down during transport, the resulting finer-grained sediment is transported onto and deposited on the abyssal plain forming sediment waves and sediment sheets. Of the debris avalanches or sediment flows, none can confidently be identified as of very recent origin (e.g. as formed on the 26 December 2004).

Sumatra overview

Our data indicate that mass wasting in the survey area takes the form of small-scale events; blocky debris avalanches or debris flows. This throws into perspective a widely held view that earthquakes trigger submarine failures that may create destructive tsunamis (e.g. Hampton *et al.* 1996; Lee *et al.* 2007). One control on large-scale submarine landsliding is earthquake frequency along the Sumatra convergent margin. The margin is active, resulting in frequent small magnitude events that would trigger numerous small-scale failures. Earthquake frequency would also preclude the accumulation of large sediment build-ups that would be prone to

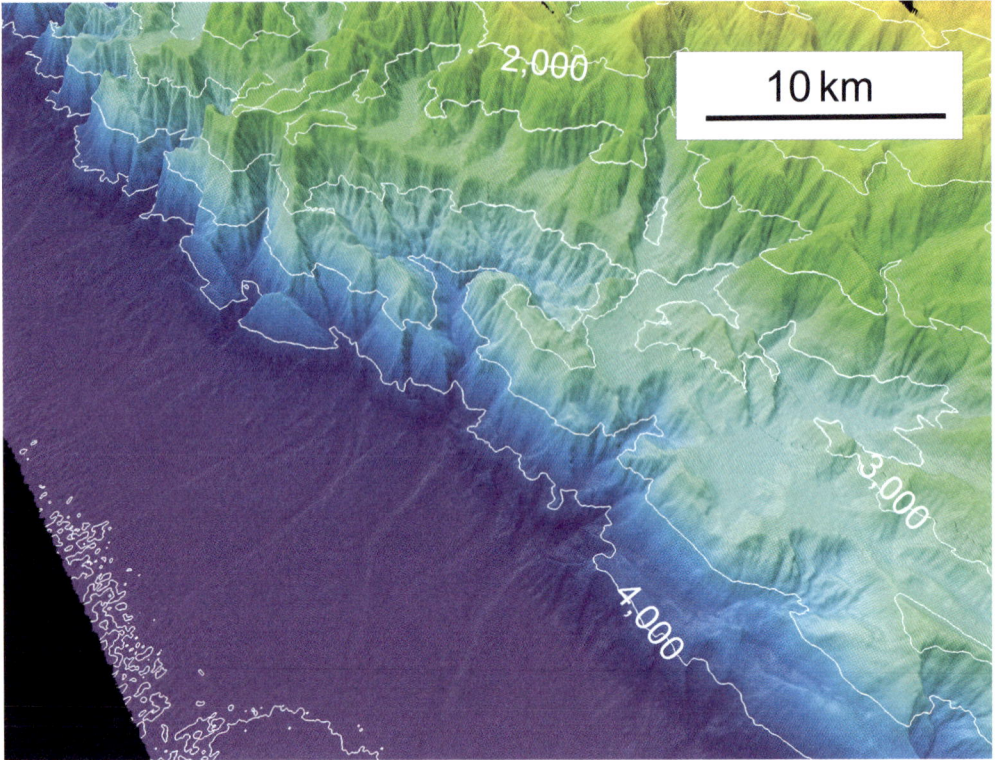

Fig. 11. The accretionary prism toe where young thrust folds are absent. Outboard of the accretionary prism toe there are linear channels, and few out-runner blocks at the base of the toe scarp. Note the heavily incised 'cauliform' morphology of erosion in the lower prism toe (for the location see Fig. 8) (vertical exaggeration ×4).

catastrophic failure. Another control on seabed failure is the overall regional framework of sediment supply.

Sediment forming accretionary prisms may be derived from the land or the subducting plate, and the morphology of the prism off of Sumatra indicates that it is sediment starved. There is no evidence for significant sediment input from the landward direction; no major canyons cross the accretionary prism. Sediment derived from the land is of small volume and trapped within the Aceh forearc basin. The interior of the prism is degraded. There also appears to be little erosion taking place on the fold limbs. The synclines between the uplifted thrust folds form elongate basins that are flat floored with little evidence of sediment fans; they are filled mainly with hemipelagic sediment. The sediment comprising the accretionary prism appears to be dominantly derived from the oceanic plate. On the prism toe, the sediment derived by mass wasting is small scale. Here there is a southwards decrease in the size of the thrust folds, an increasing isolation of the thrust folds from the prism toe and an increase in their erosion. All

indications are that there is a decrease in sediment supply to the south.

On the abyssal plain, the seismic data record a change in the sedimentation regime at some previous time (Tappin et al. 2007). There is an extensive system of channels present at depth; there are none at the surface. Channel size and internal structure record a previous period of vigorous activity that is margin parallel. Consideration of sedimentation in the Indian Ocean suggests that there should be a considerable volume of sediment delivered to the accretionary prism off Sumatra derived from the Bengal Fan (Curray et al. 2002). This does seem to be the case, but the data suggest that the reduction is greater than might be expected. The southwards increase in degeneration of the thrust ridges on the toe of the prism indicates a decrease in sediment supply in the same direction over recent timescales. This may be associated with the decrease in trench-parallel sediment supply as indicated by the lack of recent channelling. The decrease in sediment supply is attributed to the collision, during the early Quaternary, of the Nineteast Ridge with the Sunda Margin. This collision resulted in a cut-off of supply as sediment was diverted away from the margin.

Discussion and conclusions

The development of DEMs in the marine domain based on multibeam bathymetry has led to a revolution in how seabed morphology is visualized. The 3D visualization of multibeam data in interactive software such as *Fledermaus* allows seabed relief to be viewed and interpreted in a similar fashion to DEMs of topography on land. For the investigation of marine hazards this has resulted in major advances in our capability to map submarine landslides, failure of which has the potential to create destructive tsunamis. Mapping of the 3D architecture of submarine landslide failures using seabed morphology, in association with subseabed seismic data, provides the foundation for the development of the new mathematical simulations of tsunami they create (e.g. Tappin *et al.* 2008). Three-dimensional interactive DEMs of multibeam bathymetric data allow more detailed interpretations of seabed morphology than that previously possible with 2D contoured data. In addition, DEM programs such as *Fledermaus* allow integration of bathymetry with backscatter intensity maps (draped over the bathymetry) (see Figs 6b & 7b) and seismic profiles (inserted below the bathymetric surface with the seabed reflection tied to the bathymetry). Without this high degree of 3D visualization the tsunami simulations could not have been developed to the degree of accuracy now possible. The improvements in data visualization made in PNG and their impact on tsunami modelling are not unique. A comparable improvement in visualization is also seen in the research on the Storegga submarine landslide (cf. Bugge 1983; Bryn *et al.* 2005).

One of the first examples of using marine DEMs on multibeam data specifically acquired for tsunami hazard research soon after a catastrophic event was in PNG. Prior to the PNG event, most tsunami sourced from submarine landslides were modelled from hypothesized examples or based on experiments using flume studies (e.g. Jiang & LeBlond 1992, 1994). Bathymetric data were used in some cases (e.g. Murty 1979; Harbitz 1992) but not to the level of detail available from multibeam data. This led to the erroneous conclusion by some scientists that submarine landslides were not a source of catastrophic tsunami (e.g. Jiang & LeBlond 1992, 1994). The PNG tsunami proved that multibeam bathymetric data are essential not only to locate and map submarine landslides but also to provide the basis for the tsunami models. This probably was the first time that a mathematical model of a landslide tsunami was devised specifically on a marine dataset specifically acquired for the purpose. In Hawaii, DEMs of the GSLs have led to their detailed mapping that, again, forms the basis of tsunami modelling.

In the instance of Sumatra, the multibeam bathymetry has resulted in a better appreciation of the relationship between earthquakes, subduction zones, submarine sediment failure and tsunami. Although earthquakes along convergent margins trigger landslides that cause tsunami, there are other factors that need to be taken into consideration when assessing the tsunami hazard in these regions. These include margin structure, sediment supply and tectonism. The Sumatra margin is prone to large earthquakes, but is sediment starved. In addition, the margin is accretionary and there are frequent earthquakes. The result is small-scale (but probably frequent) landslides. The margin off northern PNG is also sediment starved, with frequent earthquakes. It is also undergoing rapid tectonic erosion and subsidence that extends from the prism toe to the upper (shallower water) sections of the upper trench wall. The difference between the two areas is that with PNG the location of the landslide tsunami source was proximal to the coast and was a rotational slump rather than a landslide. The major factors influencing tsunami hazard in the PNG area are a combination of the type of sediment being deposited (hemipelagic mud) and an actively subsiding margin (due to active tectonic erosion). In assessing the hazard from tsunami in convergent margin environments an appreciation of the local tectonic evolution is necessary to identify the major controls on the sedimentary regime. DEMs from multibeam bathymetry provide a fundamental basis for these assessments.

With regard to the understanding of volcanic flank collapse, a historical review of research on the GSLs in Hawaii shows how the changing nature of the data – from the acquisition of first single-beam, then sidescan backscatter data and, finally, multibeam bathymetric data – affected the evolution of ideas on the GSLs. Ultimately, it was DEMs based on the multibeam depth data that provided the detailed seabed morphology which allowed the reconstruction of the large block GSLs. These DEMs formed the basis of the tsunami simulations.

For the Sumatra data the author thanks the ship's company of HMS *Scott* for acquiring the multibeam data and the research co-ordination by the Joint Environment Directorate of Defence Intelligence, The Royal Navy, United Kingdom Hydrographic Office and the Government of Indonesia. For the PNG data the author thanks the Government of Japan for funding the marine surveys, and the Japan Marine Science and Technology Centre for the organization and provision of ships and personnel. Multibeam bathymetry in Figures 2b and 4–8 is underlain by satellite altimetry (SRTN30) exported as .cd files from the SIO website (Smith & Sandwell 1997). Thanks to D. Long and S. Day for their full and constructive reviews. D. R. Tappin publishes by permission of the Executive Director of the British Geological Survey (NERC). Mention of trade names is for identification purposes only and does not constitute endorsement.

References

ABE, K. 1979. Size of great earthquakes of 1837–1974 inferred from tsunami data. *Journal of Geophysical Research*, **84**, 1561–1568.

AIDA, I. 1975. Numerical experiments of the tsunami associated with the collapse of Mount Maeyama in 1792. *Journal of Seismological Society of Japan*, **28**, 449–460.

BRYN, P., BERG, K., FORSBERG, C. Г., SOLHEIM, A. & KVALSTAD, T. J. 2005. Explaining the Storegga Slide. *Marine and Petroleum Geology*, **22**, 11–19.

BUGGE, T. 1983. *Submarine Slides on the Norwegian Continental Margin, with Special Emphasis on the Storegga Area*. Continental Shelf Institute Publications, **110**.

CURRAY, J. R., EMMEL, F. J. & MOORE, D. G. 2002. The Bengal Fan: morphology, geometry, stratigraphy, history and processes. *Marine and Petroleum Geology*, **19**, 1191–1223.

HAMPTON, M. A., LEE, H. J. & LOCAT, J. 1996. Submarine landslides. *Reviews of Geophysics*, **34**, 33–59.

HARBITZ, C. B. 1992. Model simulations of tsunamis generated by the Storegga slides. *Marine Geology*, **105**, 1–21.

HENSTOCK, T., MCNEILL, L. C. & TAPPIN, D. R. 2006. Seafloor morphology of the 26 December 2004 Indian Ocean earthquake rupture zone. *Geology*, **34**, 485–488, doi: 10.1130/22426.1.

HUGHES CLARKE, J. E., MAYER, L. A. & WELLS, D. E. 1996. Shallow-water imaging multibeam sonars: a new tool for investigating seafloor processes in the coastal zone and on the continental shelf. *Marine Geophysical Researches*, **18**, 607–629.

JIANG, L. & LEBLOND, P. H. 1992. The coupling of a submarine slide and the surface wave which it generates. *Journal of Geophysical Research*, **97**, 12 731–12 744.

JIANG, L. & LEBLOND, P. H. 1994. Three dimensional modelling of tsunami generation due to submarine mudslide. *Journal of Physical Oceanography*, **24**, 559–573.

JOHNSON, R. W. 1987. Large-scale volcanic cone collapse: the 1888 slope failure of Ritter Volcano, and other examples from Papua New Guinea. *Bulletin of Volcanology*, **49**, 669–679.

JOHNSON, J. M. & SATAKE, K. 1997. Estimation of seismic moment and slip distribution of the April 1, 1946 Aleutian tsunami earthquake. *Journal of Geophysical Research*, **102**, 11 765–11 774.

KANAMATSU, T., HERRERO-BERVERA, E. & MCMURTRY, G. M. 2002. Magnetostratigraphy of deep-sea sediments from piston cores adjacent to the Hawaiian Islands: implication for ages of turbidites derived from submarine landslides. *In*: TAKAHASHI, E., LIPMAN, P. W., GARCIA, M. O., NAKA, J. & ARAMAKI, S. (eds) *Hawaiian Volcanoes: Deep*

Underwater Perspectives. American Geophysical Union Geophysical Monograph, **128**, 51–64.

KAWATA, Y., BENSON, B. C., BORRERO, J., DAVIES, H., DE LANGE, W., IMAMURA, F. & SYNOLAKIS, C. E. 1999. Tsunami in Papua New Guinea was as intense as first thought. *Eos, Transactions American Geophysical Union*, **80**, 101, 104–105.

LEE, H. J., LOCAT, J. ET AL. 2007. Submarine mass movements. *In*: NITTROUER, C. A. (ed.) *Continental-Margin Sedimentation: Transport to Sequence*. International Association of Sedimentologists, Special Publications, **37**, 213–274.

LIPMAN, P. W., NORMARK, W. R., MOORE, J. G., WILSON, J. B. & GUTMACHER, C. 1988. The giant submarine Alika debris slide, Mauna Loa, Hawaii. *Journal of Geophysical Research*, **93**, 4279–4299.

MCMURTRY, G. M., FRYER, G. J. ET AL. 2004a. Megatsunami Deposits on Kohala Volcano, Hawaii from Flank Collapse of Mauna Loa. *Geology*, **32**, 741 744.

MCMURTRY, G. M., WATTS, P., FRYER, G. J., SMITH, J. R. & IMAMURA, F. 2004b. Giant landslides, megatsunamis, and paleo-sea level in the Hawaiian Islands. *Marine Geology*, **203**, 219–233.

MOORE, G. W. & MOORE, J. G. 1988. Large-scale bedforms in boulder gravel produced by giant waves in Hawaii. *In*: CLIFTON, H. E. (ed.) *Sedimentologic Consequences of Convulsive Geologic Events*. Geological Society of America, Special Paper, **229**, 101–110.

MOORE, J. G. 1964. *Giant Submarine Landslides on the Hawaiian Ridge*. US Geological Survey Professional Paper, **501**, D95–D98.

MOORE, J. G. & CLAGUE, D. A. 2002. Mapping of the Nuuanu and Wailau Landslides in Hawaii. *In*: TAKAHASHI, E., LIPMAN, P. W., GARCIA, M. O., NAKA, J. & ARAMAKI, S. (eds) *Hawaiian Volcanoes: Deep Underwater Perspectives*. American Geophysical Union Geophysical Monograph, **128**, 223–244.

MOORE, J. G. & MOORE, G. W. 1984. Deposit from a giant wave on the island of Lanai, Hawaii. *Science*, **226**, 1312–1315.

MOORE, J. G., NORMARK, W. R. & HOLCOMB, R. T. 1994. Giant Hawaiian underwater landslides. *Science*, **264**, 46–47.

MOORE, J. G., CLAGUE, D. A., HOLCOMB, R. T., LIPMAN, P. W., NORMARK, W. R. & TORRESAN, M. E. 1989. Prodigious submarine landslides on the Hawaiian Ridge. *Journal of Geophysical Research*, **94**, 17 465–17 484.

MORAN, K. & TAPPIN, D. 2006. *SEATOS 2005 Cruise Report: Sumatra Earthquake and Tsunami Off Shore Survey (SEATOS)*. Available online at: http://ocean.oce.uri.edu/seatos.

MURTY, T. S. 1979. Submarine slide-generated water waves in Kitimat Inlet, British Columbia. *Journal of Geophysical Research*, **84**, 7777–7779.

RUBIN, K. H., FLETCHER, C. H. III & SHERMAN, C. 2000. Fossiliferous Lanai deposits formed by multiple events rather than a single giant tsunami. *Nature*, **408**, 675–681.

SATAKE, K. 2007. Volcanic origin of the 1741 Oshima-Oshima tsunami in the Japan Sea. *Earth Planets Space*, **59**, 381–390.

SATAKE, K. & KATO, Y. 2001. The 1741 Oshima-Oshima eruption: extent and volume of submarine debris avalanche. *Geophysical Research Letters*, **28**, 427–430.

SATAKE, K., SMITH, J. R. & SHINOZAKI, K. 2002. Three-dimensional reconstruction and tsunami model of the Nuuanu and Wailau Giant Landslides, Hawaii. *In*: TAKAHASHI, E., LIPMAN, P. W., GARCIA, M. O., NAKA, J. & ARAMAKI, S. (eds) *Hawaiian Volcanoes: Deep Underwater Perspectives*. American Geophysical Union Geophysical Monograph, **128**, 333–346.

SCHWAB, W. C., DANFORTH, W. W., SCANLON, K. M. & MASSON, D. G. 1991. A giant submarine slope failure on the northern insular slope of Puerto Rico. *Marine Geology*, **96**, 237–246.

SHINOZAKI, K., REN, Z.-Y. & TAKAHASHI, E. 2002. Geochemical and petrological characteristics of Nuuanu and Wailau Landslide Blocks. *In*: TAKAHASHI, E., LIPMAN, P. W., GARCIA, M. O., NAKA, J. & ARAMAKI, S. (eds) *Hawaiian Volcanoes: Deep Underwater Perspectives*. American Geophysical Union Geophysical Monograph, **128**, 297–310.

SMITH, J. R., SATAKE, K., MORGAN, J. K. & LIPMAN, P. W. 2002. Submarine landslides and volcanic features on Kohala and Mauna Kea volcanoes and the Hana Ridge, Hawaii. *In*: TAKAHASHI, E., LIPMAN, P. W., GARCIA, M. O., NAKA, J. & ARAMAKI, S. (eds) *Hawaiian Volcanoes: Deep Underwater Perspectives*. American Geophysical Union Geophysical Monograph, **128**, 11–28.

SMITH, M. S. & SHEPHERD, J. B. 1996. Tsunami waves generated by volcanic landslides: an assessment of the hazard associated with Kick 'em Jenny. *In*: MCGUIRE, W. J., JONES, A. P. & NEUBERG, J. (eds) *Volcano Instability on the Earth and Terrestrial Planets*. Geological Society, London, Special Publications, **110**, 115–123.

SMITH, W. H. F. & SANDWELL, D. T. 1997. Global seafloor topography from satellite altimetry and ship depth soundings. *Science*, **277**, 1957–1962.

STEARNS, H. T. 1978. Quaternary shorelines in the Hawaiian Islands. *Bernice P. Bishop Museum Bulletin*, **237**, 57.

STEARNS, H. T. & CLARK, W. O. 1930. *Geology and Water Resources of the Kau District, Hawaii*. US Geological Survey Water-Supply Paper, **616**, 1–94.

STEARNS, H. T. & MACDONALD, G. A. 1946. Geology and groundwater resources of the Island of Hawaii. *Hawaii Division of Hydrography Bulletin*, **9**, 363.

SWEET, S. & SILVER, E. A. 2003. Tectonics and slumping in the source region of the 1998 Papua New Guinea tsunami from seismic reflection images. *Pure and Applied Geophysics*, **160**, 1945–1968, doi: 10.1007/s00024–003–2415-z.

TAKAHASHI, E., LIPMAN, P. W., GARCIA, M. O., NAKA, J. & ARAMAKI, S. (eds) 2002. *Hawaiian Volcanoes: Deep Underwater Perspectives*. American Geophysical Union Geophysical Monograph, **128**.

TAPPIN, D. R., MATSUMOTO, T. *ET AL.* 1999. Sediment slump likely caused 1998 Papua New Guinea tsunami. *Eos, Transactions of the American Geophysical Union*, **80**, 329, 334, 340.

TAPPIN, D. R., WATTS, P., MCMURTRY, G. M., LAFOY, Y. & MATSUMOTO, T. 2001. The Sissano, Papua New Guinea tsunami of July 1998 – offshore evidence on the source. *Marine Geology*, **175**, 1–23.

TAPPIN, D. R., WATTS, P., MCMURTRY, G. M., LAFOY, Y. & MATSUMOTO, T. 2002. Prediction of slump generated tsunamis: the July 17th PNG event. *Science of Tsunami Hazards*, **20**, 228–238.

TAPPIN, D. R., WATTS, P. & MATSUMOTO, T. 2003. Architecture and failure mechanism of the offshore slump responsible for the 1998 PNG tsunami. *In*: LOCAT, J. & MEINERT, J. (eds) *Submarine Mass Movements and their Consequences*. Kluwer, Dordrecht, 383–392.

TAPPIN, D. R., MCNEIL, L. C., HENSTOCK, T. & MOSHER, D. 2007. Mass wasting processes – offshore Sumatra. *In*: LIKOUSIS, V. & SAKELLARIOU, D. (eds) *Submarine Landslides*. Springer, Dordrecht.

TAPPIN, D. R., WATTS, P. & GRILLI, S. 2008. The Papua New Guinea tsunami of 17 July 1998: anatomy of a catastrophic event. *Natural Hazards and Earth System Science*, **8**, 243–266.

WARD, S. N. & DAY, S. 2003. Ritter Island volcano-lateral collapse and the tsunami of 1888. *Geophysical Journal International*, **154**, 891–902.

WIEGEL, R. L. 1964. *Oceanographical Engineering*. Prentice-Hall, Englewood Cliffs, NJ.

LiDAR basics for natural resource mapping applications

JAMES D. GIGLIERANO

*Iowa Geological Survey, Iowa Department of Natural Resources, 109 Trowbridge Hall,
Iowa City, IA 52242-1319, USA (e-mail: james.giglierano@dnr.iowa.gov)*

Abstract: LiDAR elevation data is becoming widely available for use by many non-engineering
mapping specialists such as geologists, soil scientists and planners. Understanding the basics of
LiDAR data acquisition is essential to using the data effectively in mapping applications, including
how vegetation affects the vertical accuracy of LiDAR. Tools are available for mapping specialists
to process raw LiDAR data into useful GIS products so they do not have to rely on vendor supplied
products.

The purpose of this paper is to help newcomers
understand the basics of Light Detection And
Ranging (LiDAR) data collection and processing,
especially non-engineering, mapping specialists
such as geologists, soil scientists and those in interested in land cover characterization. LiDAR is being
increasingly used worldwide for the collection of
detailed elevation data. In the USA many states
are embarking on large-scale LiDAR acquisitions,
and inevitably LiDAR elevation and other derived
products will become widely available to many
different audiences. To make full use of this new
source of information a knowledge of LiDAR data
collection and handling procedures will be required,
as well as guidance for the conversion and utiliz-
ation of vendor-supplied files. In some cases
mappers may have to perform the processing themselves, or may ask for this to be done by vendors or
third parties. In other cases, LiDAR-derived topographical data may be supplied by a local government entity that has no metadata, and in these
instances the user will have to make some assumptions about the type of processing that may have
been performed.

How LiDAR data is collected and represented

The LiDAR equipment basically consists of a laser
rangefinder operating from some form of airborne
platform (helicopter, plane or satellite) that makes
repeated measurements of the distance from the
platform to the ground. The position and elevation
of the platform is precisely known, using airborne
GPS along with ground control, so that the elevation
of the ground surface can be calculated by subtracting the laser rangefinder distance from the height
of the platform. Compensation must be made for
the tilt and pitch of the airborne platform using

gyroscopes and accelerometers in the aircraft's
inertial measurement unit. A good technical overview of LiDAR scanning technology is provided
by Wehr & Lohr (1999).

LiDAR systems record thousands of highly accurate distance measurements every second (newer
systems up to 150 kHz; older systems 30–80 kHz)
and create a very dense coverage of elevations over
a wide area in a short amount of time. Because
LiDAR is an active sensor that supplies its own
light source, it can be used at night thus avoiding
routine air traffic. It can also be flown under some
types of high cloud conditions. Most LiDAR
systems record multiple surface reflections, or
'returns', from a single laser pulse. When a laser
pulse encounters vegetation, power lines or buildings, multiple returns can be recorded. The first
return will represent the elevation near the top
of the object. Second and third returns may
represent trunks and branches within a tree, or
understorey vegetation. Hopefully, the last return
recorded by the sensor will be the remaining laser
energy reflecting off the ground surface, but sometimes the tree blocks all the energy from reaching
the ground. These multiple returns can be used to
determine the height of trees or power lines, or
give indications of forest structure (crown height,
understorey density, etc.). Figure 1 shows a single
2×2 km tile consisting of 3.3 million first return
LiDAR points.

Another feature of an airborne LiDAR system
is the use of mirrors or other technology to point
the laser beam to either side of the aircraft as
it moves along its flight path. Depending on the
scanning mechanism, the LiDAR scans can have a
side-to-side, zig-zag, sinusoidal or wavy pattern.
While the laser itself pulses many thousands
of times per second, the scanning mechanism
usually moves from side to side at around 20–
40 cycles s^{-1}. This scanning, combined with the

From: FLEMING, C., MARSH, S. H. & GILES, J. R. A. (eds) *Elevation Models for Geoscience.*
Geological Society, London, Special Publications, **345**, 103–115.
DOI: 10.1144/SP345.11   0305-8719/10/$15.00 © The Geological Society of London 2010.

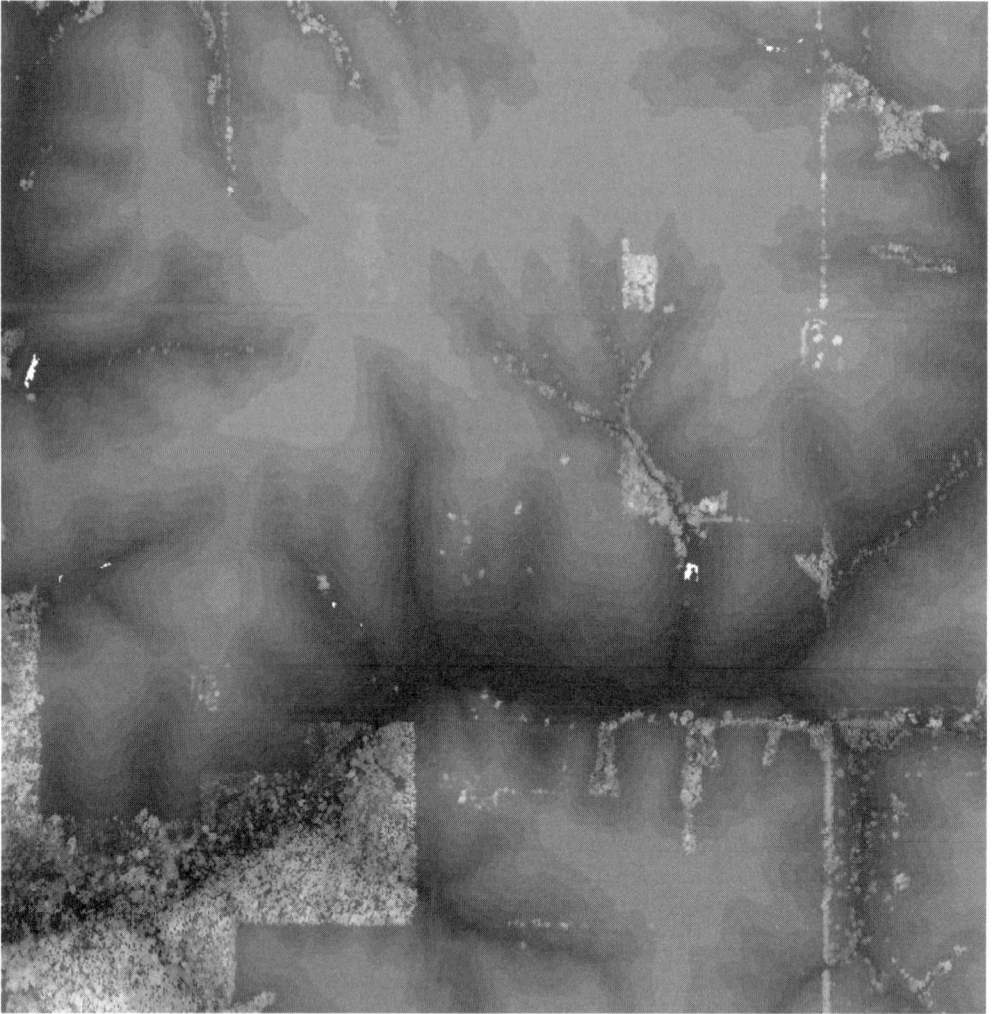

Fig. 1. Grey scale image consisting of 3.3 million LiDAR first return LiDAR points. First returns indicate the tops of trees and buildings, as well as bare ground in open areas. White areas are data voids where no returns were recorded, usually from non-reflecting water surfaces.

forward motion of the aircraft, produces millions of elevations in a short distance and time. The field of view or angle the scan makes from side to side can be adjusted by the operator, but usually is $30°-40°$. This creates a swathe of around 1 km in width or less. Adjacent swathes overlap by 15–30% so that no data gaps are left between flight lines.

The spacing of LiDAR points on the ground is a function of the laser pulse frequency, scan frequency and flight height (Baltsavias 1999). While there is usually a nominal or average point spacing specified in a LiDAR project, actual data points have variable spacings that are smaller and larger

than the specified spacing. Mappers need to be aware of these effects when viewing final products that were derived from the raw data. The second aspect is that because the laser scans from side to side, it interacts with the ground in different ways depending on the angle of incidence. LiDAR pulses at the edge of a scan will strike the sides of buildings, whereas pulses at the centre of a scan will only strike the tops of roofs. Likewise, pulses at the edges of scans will pass through trees at an angle. Sometimes this will create 'shadows' on the other side where no LiDAR passes through. In addition, less energy will return to the LiDAR receiver as it reflects away from the aircraft. This is evident in the images created from the intensity values for each return: one can see overall darkening of the intensity at the edges of swathes. The latter appear darker than the returns at the centre of swathes.

How LiDAR points are processed into TINs and DEMs

In the spring of 2005 the Iowa Department of Natural Resources (DNR) and others acquired LiDAR coverage with a nominal resolution of 1 m over the Lake Darling watershed from a commercial vendor. The vendor's LiDAR system collected a first and last return from each LiDAR pulse. From the first and last returns a so-called 'bare earth' return was created using a proprietary classification algorithm developed by the vendor. Such classification systems try to sort out non-bare earth returns (tree tops, buildings, power lines, vehicles) from bare earth (ground surface) returns. They use differences in elevation between the first and last returns, relative changes in elevation and slope to distinguish bare earth in forested areas. Intensity data are used to identify vegetation and man-made materials. The LiDAR data for the Lake Darling watershed was collected in April, before most trees and bushes had leaves. There are some data voids in forested areas owing to non-penetration of the laser through the tree canopy, but these areas are generally less than 10 m across and are easily filled in by interpolation. Leaf-on conditions and tall crops, such as corn, do not allow easy penetration of the laser beam to the ground and should be avoided.

LiDAR data for the Lake Darling area came from the vendor in the form of 2 × 2 km tiles with x and y co-ordinates, z elevations and intensity values in ASCII text format. With a nominal 1 m posting spacing, some tiles had up to 3.3 million points. Points near the centre of the flight lines were close to the nominal 1 m spacing (Fig. 2), while towards the ends of scans the points converge with the start of the next scan (Fig. 3). In this dataset the scanning pattern was a zig-zag, which made some points converge while others diverged. The points can be less than half of the nominal spacing and, likewise, where they diverge they can be twice the nominal spacing. Because some points can be as close as 0.5 m, the tiles were initially interpolated to create grids with 0.5 m resolution, the idea being that no data points should be merged or averaged with any other points. There is a tendency among some users to create grids with resolutions of 3, 5 and even 10 m in order to save storage space, or as a way to reduce the volume of data to process. We

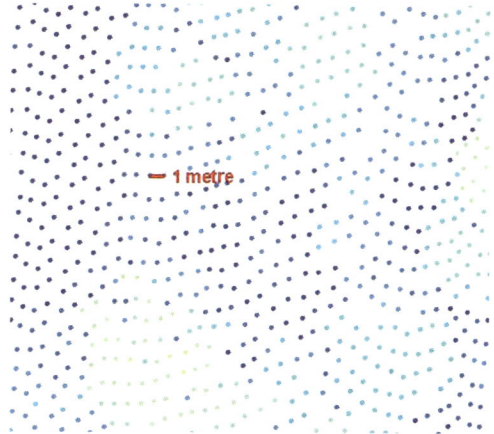

Fig. 2. An area in the middle of a LiDAR flight line. LiDAR point spacing is around 1 m at the centre of back-and-forth scans.

desired to create the grids as close as possible to the native resolution of the LiDAR data in order to fully evaluate its potential to represent the smallest surface features.

To make digital elevation models (DEM) from the tiles, the Surfer 8 software (http://www.goldensoftware.com/products/surfer/surfer.shtml) was used. This software first creates a TIN (triangulated irregular network) before it interpolates the points into a raster DEM. However, once the DEM tiles were initially put together into mosaics, it

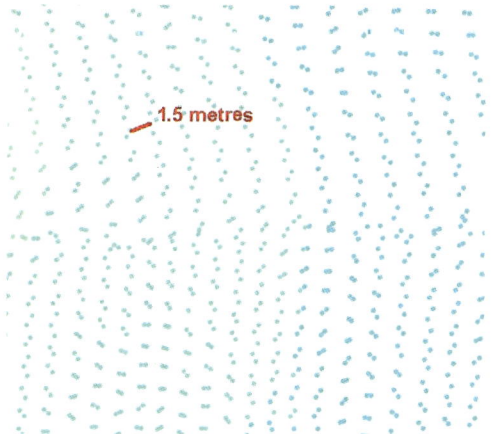

Fig. 3. An area of two adjacent LiDAR flight lines. The point spacing is highly variable at the edges of flight lines. Some points are less than 1 m apart at the end of one and beginning of the next scan, while the distance between points in different sets of scans can be as much as 1.5 m apart.

became obvious that there were noticeable gaps between each tile. To remedy this problem a short C program was created to sort through the ASCII text files of the adjacent tiles and find points within a 3 m buffer of the edge of the tile to be processed. The tiles were then reprocessed, adding the 3 m buffers, and when these raster tiles were merged together into a mosaic the gaps were almost completely eliminated. Leica Imagine (http://gi.leica-geosystems.com/) software was used to mosaic all the tiles into one large raster DEM file. From the DEM, shaded relief images were created and compressed. Digital elevation models and shaded relief images were then easily imported into ArcGIS software (http://www.esri.com/) for display and further analysis.

Field examination of the LiDAR bare earth shaded relief images was conducted in January 2006. It was surprising how well the LiDAR shaded relief images represented the smallest topographical features, including small slope changes of less than 0.5 m, even in forested areas. There were some data voids, owing to the lack of penetration through the dense tree canopy, but there were enough data points to show good definition of incised stream channels, meander scars and gullies (Fig. 4). Man-made features such as road ditches and embankments, terraces and dams were also well defined. Tillage patterns are evident as regular lineated textures on crop fields.

Because the bare earth processing does not remove 100% of the forest artefacts, a distinctive bumpy pattern remains in the bare model that indicates the presence of forest cover (Fig. 5). During field examination it was noticed that different canopy structures were represented by different patterns in the artefacts. In the tall-canopy floodplain forest most of the bumps were removed leaving a predominantly smooth surface, whereas on side slopes with a thick understorey or brush cover the texture on the shaded relief image is rougher in appearance. Interestingly, the bare earth processing removed nearly all of the numerous tree falls in the stream channels, which allows drainage tracing programs to work well when following flow paths downstream. Also, areas with pine trees were very smooth, indicating nearly complete penetration of the laser beam.

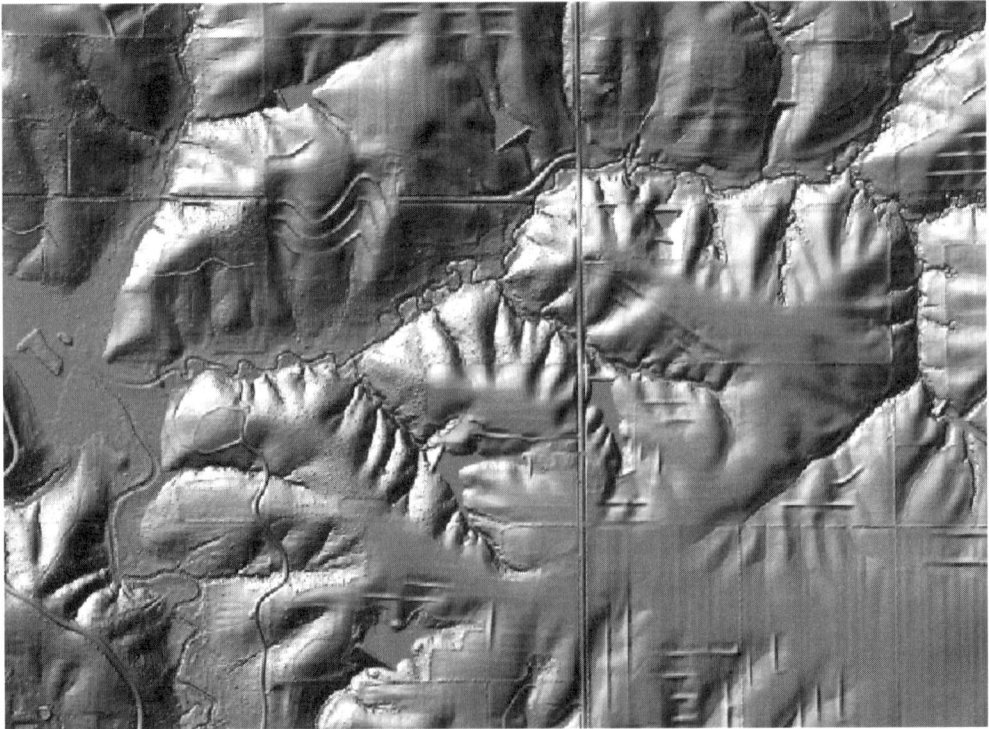

Fig. 4. Portion of the bare earth shaded relief image of the Lake Darling watershed showing natural and man-made features readily apparent in the LiDAR data. Natural and man-made drainage features, roadbeds, fence lines and tillage patterns are readily visible.

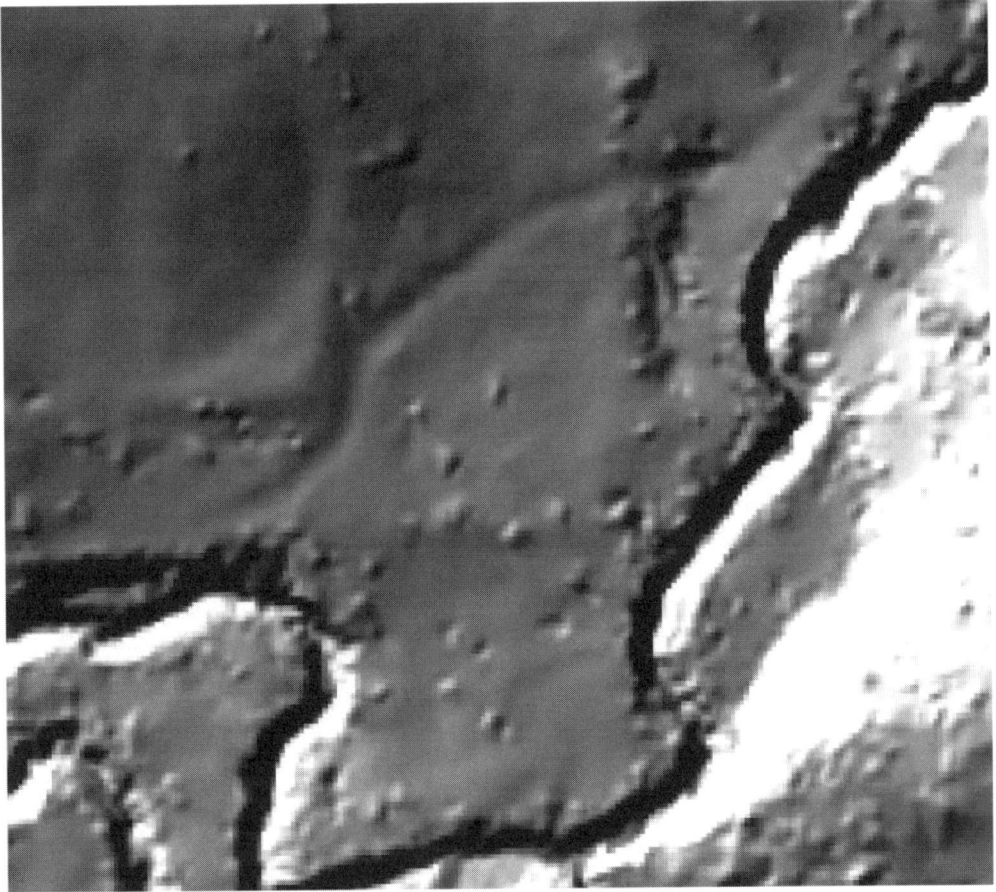

Fig. 5. Portion of a bare earth shaded relief image showing artefacts (bumpy texture) in deciduous forest areas. These artefacts are LiDAR elevations classified as bare earth, but probably are from tree trunks, branches or understorey close to the ground and classified as bare earth by the vendor's algorithm.

How to use LiDAR products for mapping applications

Once the raw LiDAR point tiles are processed into high-resolution DEMs, other useful mapping products can be derived. The derived shaded relief image previously mentioned (Fig. 4) is very useful for visual display and interpretation, and can be combined with colourized elevation images for extra information content. Another useful display product is the slope map, which can be derived from the DEM using the grid processing tools found in almost every GIS package. Usually a choice can be made as to whether the slope rate is calculated in degrees or as percentages (45° slope = 100%). A slope map based on percentage can be grouped into the slope classes typically used by soil survey mappers (A slope class = 0–2%, B = 2–4%, etc.), and can be readily compared

to soil polygons displayed by slope class (Fig. 6). Figure 7 shows the new level of detail available in slope classes derived from LiDAR data.

Besides the elevation component of the LiDAR return, many systems produce an intensity component that indicates the strength of the LiDAR return. This intensity value is mostly influenced by the reflectance of the material struck by the laser pulse, but is also influenced by the scan angle (laser pulses directed away from the airplane at significant angles do not reflect back as much light energy as a pulse directed straight down from the plane). Because most LiDAR systems use a laser that emits light in the near-infrared portion of the spectrum (LiDAR used for Lake Darling had a wavelength of 1064 nm), the intensity of LiDAR return is directly related to the near-infrared reflectance of the target material. An image constructed from the intensity component of the returns looks

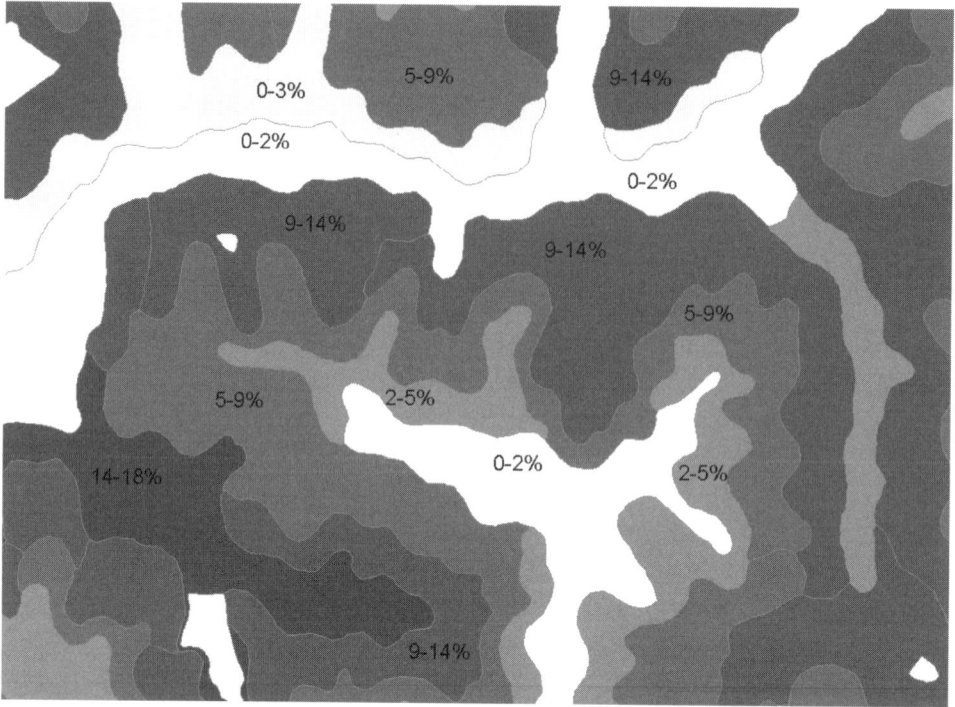

Fig. 6. Soil survey soil polygons shaded by slope class range: light shades are lower slopes and darker shades indicate steeper slopes.

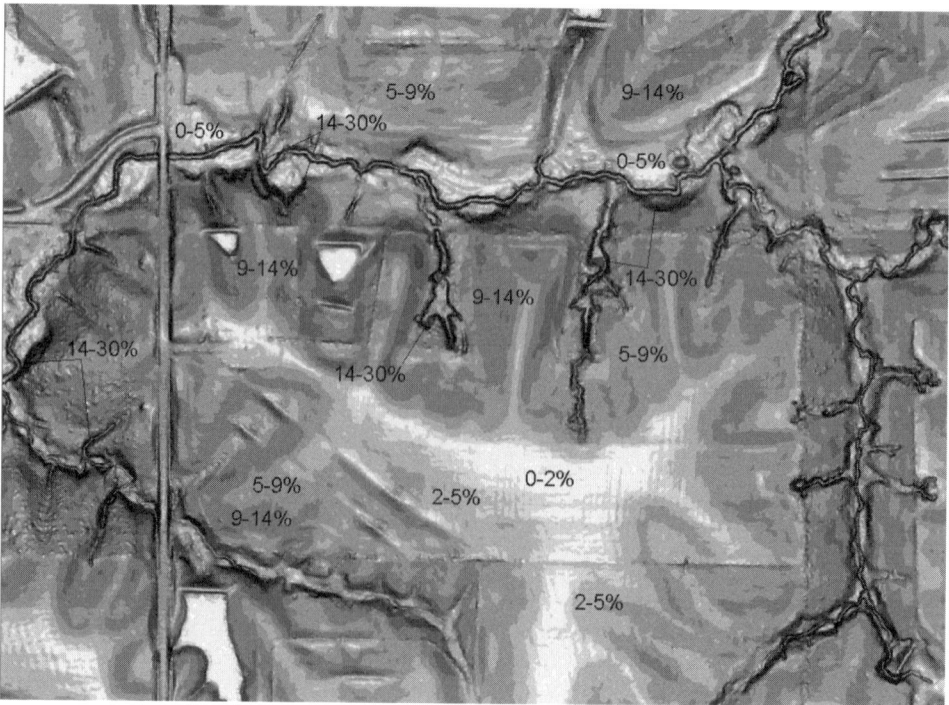

Fig. 7. Slope class ranges derived from Lake Darling LiDAR data. While low slope areas on LiDAR look similar to the soil polygons, LiDAR more clearly represents steep slopes such as gullies and stream channels.

Fig. 8. Portion of a LiDAR intensity image of the Lake Darling watershed, constructed from bare earth return intensity values.

very much like a black and white near-infrared aerial photograph (Fig. 8). An intensity image has one interesting peculiarity, however: tree shadows point away from the flight lines, so one can see shadows pointing in opposite directions close together at the edge of two flight lines. Because intensity is recorded from each LiDAR return, it is possible to construct first return intensity images as well as last return intensity images, and have them look quite different, especially in forested areas where the first return might represent mainly the tree tops but the last return intensity could represent many other features, including the forest floor.

Vertical accuracy test and influence of land cover

Usually one of the first questions asked by new LiDAR users concerns the vertical accuracy of the elevation data. In the Lake Darling project the stated accuracy is 15 cm (0.5 ft) RMSE (root mean square error) in the bare earth areas and 37 cm (1 ft) in vegetated areas. Because there are no high

accuracy geodetic monuments in the watershed, and we did not have access to survey grade GPS equipment, we needed some other way to test the vertical accuracy. Fortunately, a digital terrain model and associated 2 ft contours produced by aerial photography and photogrammetric techniques for a road project was available from the Washington County Engineer's office. The digital terrain model and contours were created by a local aerial photography firm, and had a stated vertical accuracy of 6.1 cm (0.2 ft). The area covered by the model is over 3.2 km (2 miles) long and 0.4 km (0.25 mile) wide. The digital terrain model consisted of elevation points and break lines (Fig. 9) in CAD format. Using the 3D_ANALYST extension in ArcGIS, the photogrammetrically derived terrain model was converted into a triangulated irregular network or TIN, and interpolated into a 1 m elevation grid. The LiDAR elevation grid was then subtracted from the photogrammetry grid to produce a simple difference grid: the overall average difference between the two grids was only 3.3 cm (0.11 ft). In order to compare the two grids to their stated accuracies, the RMSE had to be calculated. First, the simple difference grid was

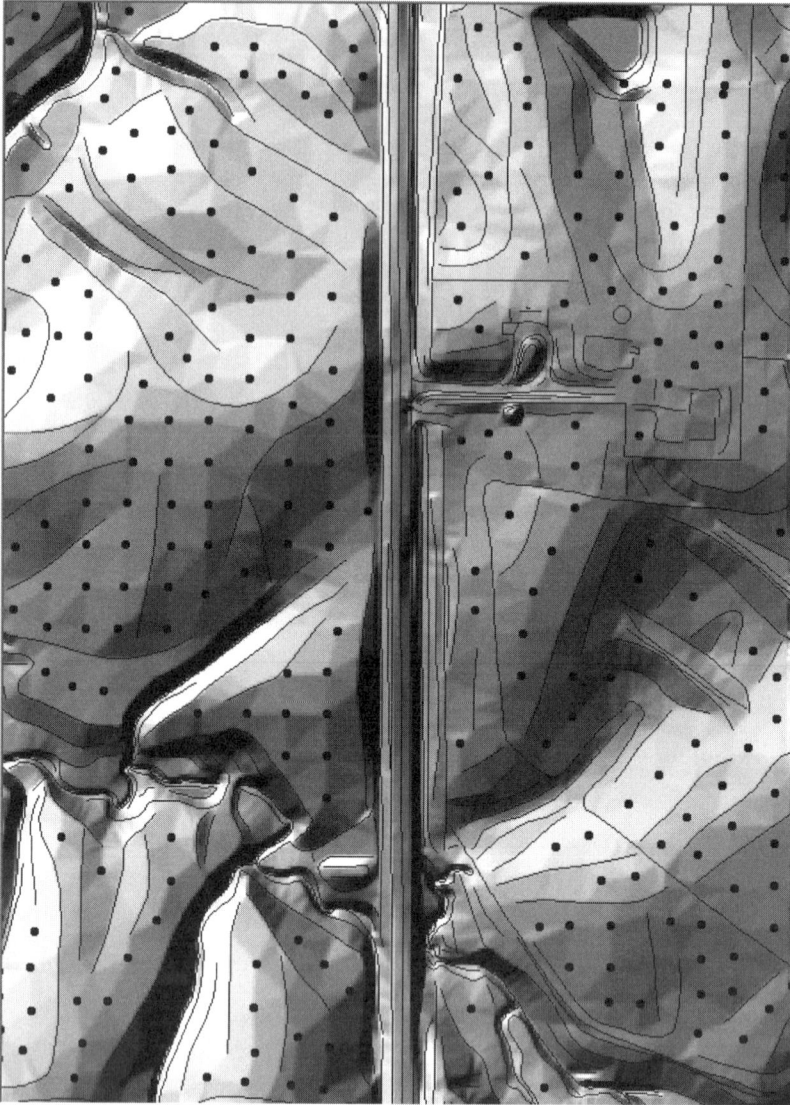

Fig. 9. A 0.4 × 0.6 km portion of shaded relief of a digital terrain model derived from low-altitude aerial photographs. The black dots are elevation mass points and the black lines are break lines.

multiplied by itself to create the squared difference grid. Using a polygon coverage of land cover from 2005, the mean squared difference was calculated for each land cover class using the zonal statistics command in ArcToolBox. By using the spatial calculator function in the SPATIAL_ANALYST extension, the square root of the values in the 'mean' field of the table, the RMSE, was found for each land cover class. The zonal statistic tool also computes a 'count' of cells for each class and a 'sum' of the elevations within that class. Calculating

the sum of all the 'count' field values and 'sum' field values for all the classes, and dividing the total sum by the total count found the average squared difference for the entire dataset. By taking the square root of this value, the RMSE was found for the whole area. Initially, RMSE between the LiDAR DEM and the photogrammetry DEM was found to be 0.79 ft or 24.1 cm.

Upon examination of the squared difference image, it was apparent that the terrain in several areas had changed significantly between the time

OID	LAND_USE	ZONE_C	COUNT	AREA	MIN	MAX	RANGE	MEAN	STD	SUM	RMSE_ft	RMSE_CM
0	Residential	1	36665	36665	0.00000	18.04	18.04	0.262028	0.823366	9607.27	0.511887	15.6023
1	Water	2	26069	26069	0.00000	10.053	10.053	0.234739	0.848544	6119.4	0.484499	14.7675
2	Pasture	3	152359	152359	0	41.5079	41.5079	0.383737	1.24089	58465.8	0.619465	18.8813
3	CRP	4	160257	160257	0	38.4666	38.4666	0.252299	0.855599	40432.7	0.502294	15.3099
4	Timber	5	121869	121869	0	106.069	106.069	0.71515	2.79821	87154.6	0.845665	25.7759
5	Wildlife/Wooded	6	1825	1825	0.00001	7.3607	7.36069	0.965743	1.03896	1762.48	0.982722	29.9534
6	Road	7	44609	44609	0	107.915	107.915	0.316614	0.942917	14123.9	0.562685	17.1506
7	Row Crop	8	353980	353980	0	328.666	328.666	0.214536	0.903116	75941.3	0.463180	14.1177
8	Alfalfa	9	9979	9979	0	8.07532	8.07532	0.179604	0.476496	1792.27	0.423797	12.9173

TOTALS 907612   295399

Mean Squ. Difference = 295399/907612 = .3255

Square root of MSD = .5705

RMSE = .57' or 17.4 cm

Fig. 10. Root mean square error (RMSE) calculation of photogrammetrically derived DEM and LiDAR DEM, after 2000–2005 change areas have been removed from the calculation. Area field is m^2.

of the aerial photography flight in 2000 and the LiDAR flight in 2005. These changes mainly included areas where the installation of sediment retention structures and dams, and road grading, had occurred. When these areas were digitized and excluded from the squared difference calculation, the overall RMSE was found to be 0.57 ft or 17.4 cm (Fig. 10). The RMSE of the row crop area was 0.46 ft (14.3 cm), the grass areas 0.62 ft (18.9 cm) and forested areas 0.85 ft (25.8 cm).

Fig. 11. Portion of a shaded relief image made from a NED 30 m DEM. The area is from the Lake Darling watershed in Washington County, Iowa.

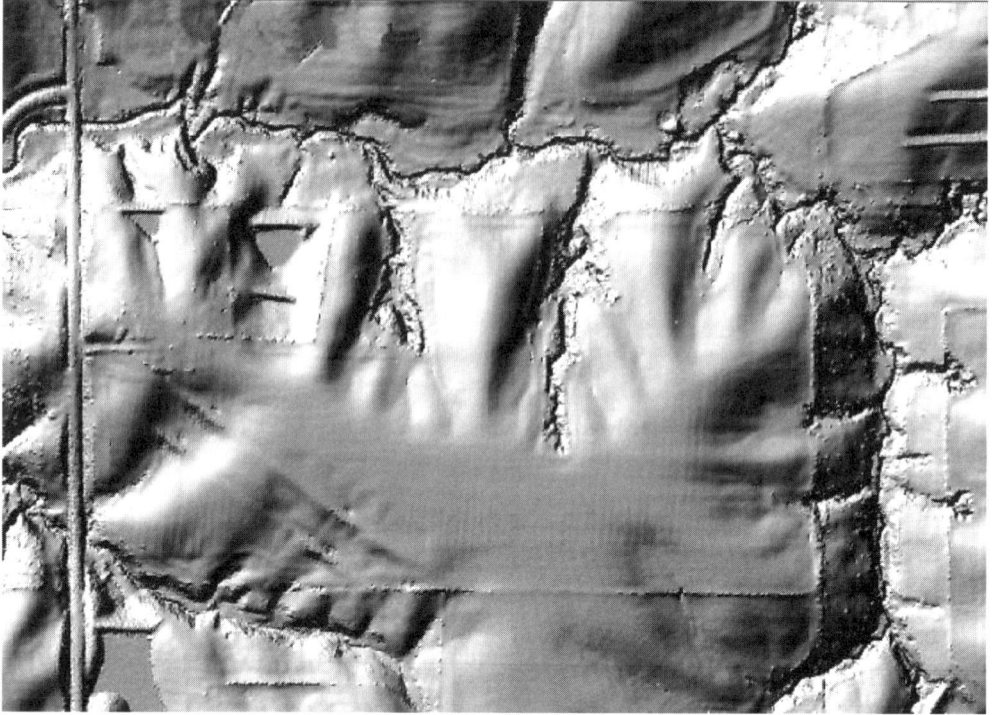

Fig. 12. Portion of a shaded relief image made from a 1 m LiDAR DEM for the same area as Figure 11 in Washington County, Iowa.

If the DEM derived by photogrammetric means is accepted as the higher accuracy source, then the LiDAR meets its stated accuracy of 15 cm in the bare ground areas and less than 37 cm in the vegetated areas. This appears to be a good test of LiDAR accuracy because it includes many types of land cover conditions, not just a few high accuracy locations at benchmarks on roads or nearby ditches.

Comparing old and new data

One of the first tests of any new LiDAR dataset is to compare it with the existing DEM derived from the 10 ft contours of the USGS (US Geological Survey) topographical quadrangle mapping projects of the latter half of the twentieth century. Displayed at smaller scales, it is difficult to see much difference between the shaded relief images derived from the 30 m resolution National Elevation Dataset or NED (http://ned.usgs.gov/) and LiDAR shaded relief. Only when the display is zoomed into larger scales is it possible to see the marked differences between the 30 m NED (Fig. 11) and LiDAR DEM (Fig. 12). On a large scale, man-made features such as roadways and ditches, fence lines, terraces

and dams are easily seen, as well as natural features such as stream channels, gullies and floodplains. None of these smaller features is discernable on the 30 m NED shaded relief image.

Where LiDAR really excels is in mapping areas with little relief. Figure 13 is a shaded relief image derived from the 30 m NED, which shows typical glaciated terrain in north-central Iowa, east of Spirit Lake in Dickinson County, Iowa. Figure 14 shows the same area using 1 m resolution LiDAR, which focuses the indistinct mounds seen on the NED shaded relief into sharply defined, circular and elongated features. These are interpreted to be the remnants of ice-walled lakes, which were formed on the surface of the last Pleistocene glacier to visit the area. These lakes filled with sediment, leaving the latter as low mounds after the ice had melted (Quade *et al.* 2004).

Figure 15 shows the Missouri River floodplain north of Council Bluffs, Iowa, in a view that again uses the 30 m resolution NED to create a shaded relief image. It reveals numerous defects in the original conversion of widely spaced contours on a very flat surface. With a 10 ft contour interval, there is not enough information to adequately interpolate features on the floodplain. For example, the

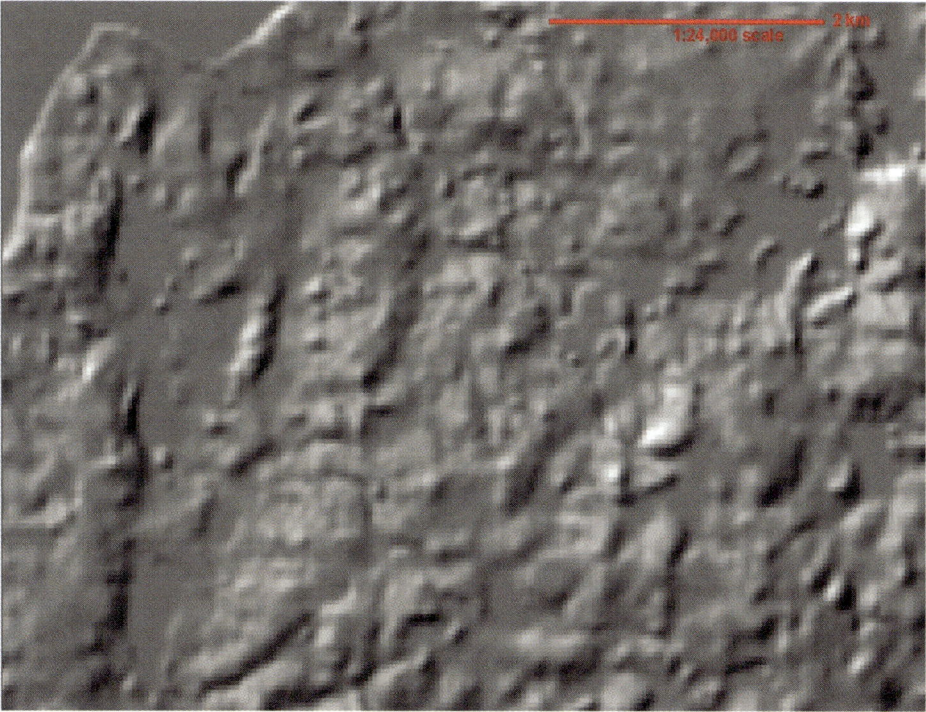

Fig. 13. Portion of a shaded relief image showing recently glaciated terrain near Spirit Lake in Dickinson County, Iowa. The shaded relief was created from a 30 m resolution DEM from the NED.

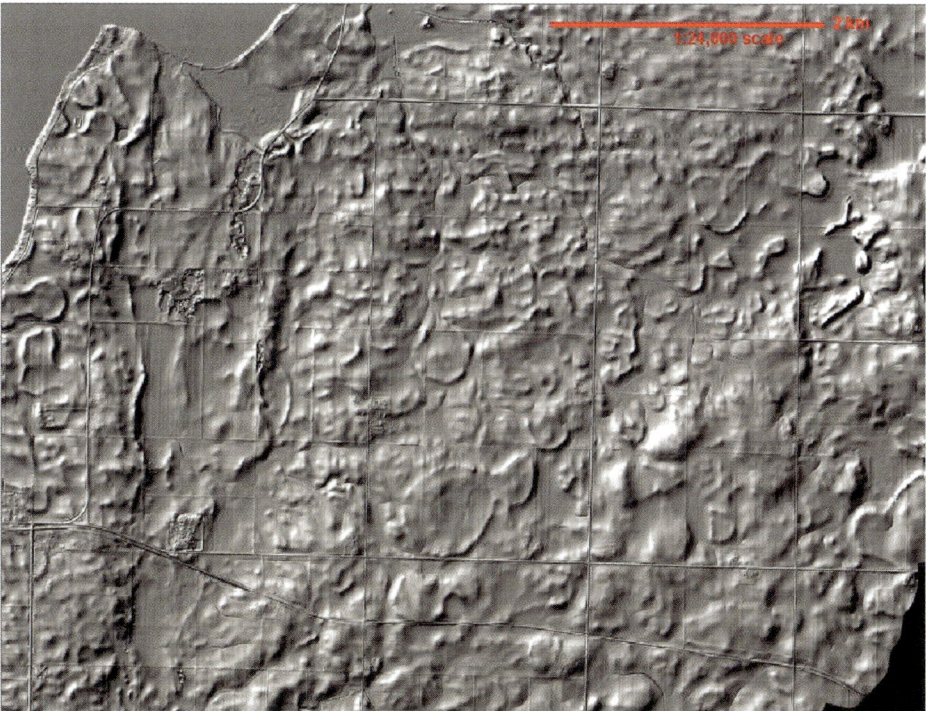

Fig. 14. Portion of a LiDAR-derived shaded relief image of the same area of glacial terrain near Spirit Lake in Dickinson County, Iowa. Notice how the shapes of subtle, low relief glacial features are now readily apparent.

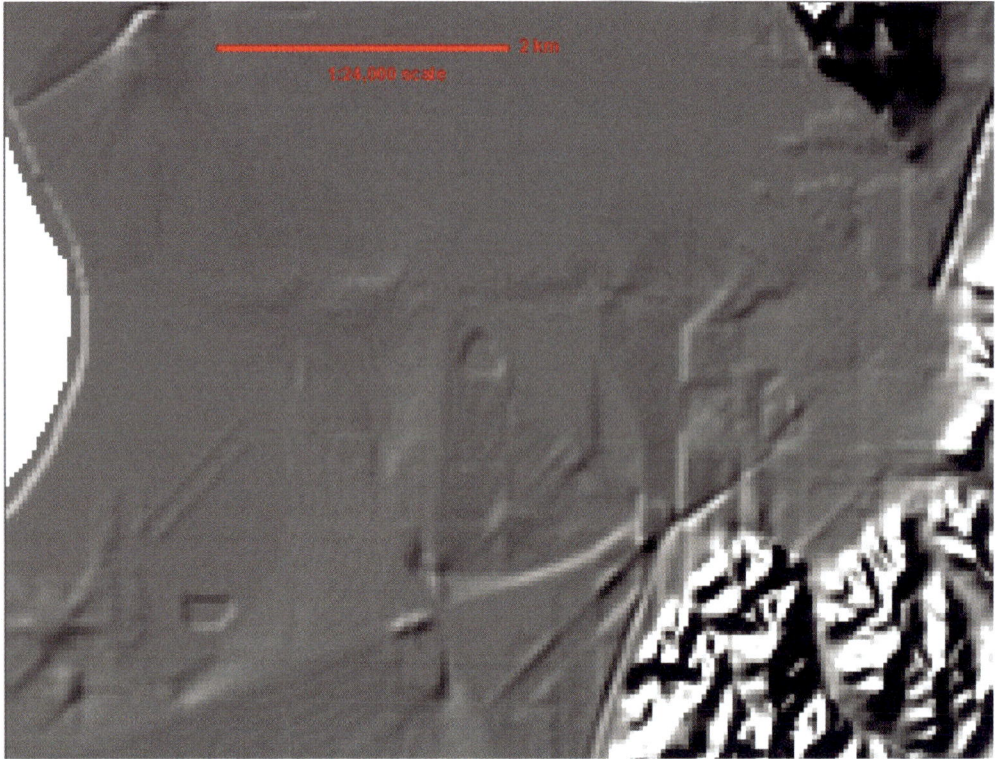

Fig. 15. Portion of a shaded relief image showing the Missouri River floodplain north of Council Bluffs, Iowa. The shaded relief image was created from a 30 m resolution DEM from the NED. Notice the cross-shaped features that are artefacts of the interpolation of the original 10 ft contours from USGS topographical maps.

shaded relief image reveals cross-shaped artefacts within the DEM, which were created by the inter- polation software trying to connect widely spaced data. Figure 16 shows the great improvement affor- ded by interpolating a surface from closely spaced LiDAR points (about 2 m LiDAR point spacing). The features revealed on the LiDAR shaded relief image include: Missouri River meander scars, levees along drainage ditches, fence lines, interstate lanes, railroad right of ways, borrow pits and sewage lagoons.

Geological mappers using shaded relief images for on-screen digitizing will need to learn new tech- niques of recognizing and separating man-made as well as geomorphic features. Because shaded relief images can represent the encoding of rela- tively small changes in slopes, mappers will need to build up criteria for the recognition of everyday features using the clues in contrast, shading, shape, texture, pattern and context contained in these images. This contrasts with past practices in which geological mappers interpreted aerial photo- graphs by poring over example after example of

natural and man-made features, and learned how to interpret geological features by looking at their geomorphic signatures on topographical maps. LiDAR will cause us to relearn and reinvent both techniques by moving the geomorphic scale down to the realm of the air photograph, roughly at resol- utions from 1 to 5 m. While qualitative information on slopes was formerly available through the use of stereo viewers and aerial photographs, only now, with the advent of LiDAR data, is there so much quantitative slope information available. With digital elevation data derived from LiDAR, new computer-assisted classification strategies can be developed for geomorphic features, as well as developing new types of imagery to support manual interpretations.

Summary

Large-scale LiDAR acquisitions will provide mapping professionals with an abundance of new high-quality elevation data to use as base maps for

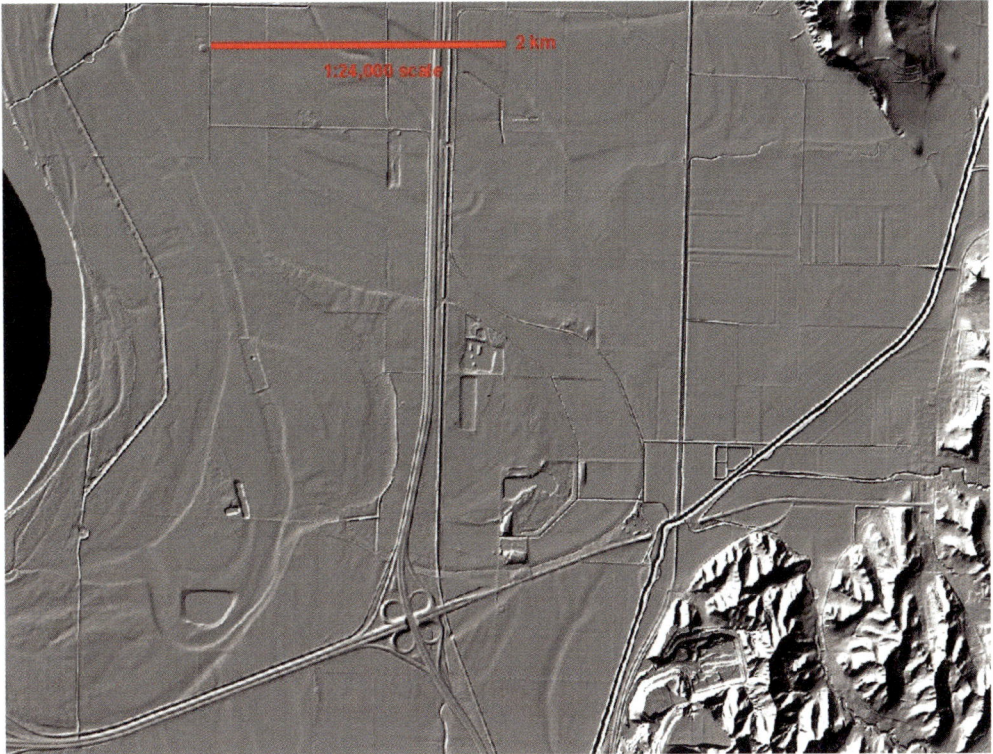

Fig. 16. Portion of a LiDAR-derived shaded relief image of the same area on the Missouri River Floodplain. Notice the much finer detail showing the interstate cloverleaf, river meander scars, borrow pits, and a ditch and levee system. Editing by the vendor removed the bridge deck. LiDAR DEM obtained from the Pottawattamie County GIS Department.

their projects. To take full advantage of this new data source, those carrying out the mapping need to be aware of how LiDAR data are collected, and the type of data reduction processes that are used by commercial vendors to make deliverable products for their clients. In many cases, mappers will want to manipulate the raw LiDAR returns into their own TINs, DEMs and derived products, but sometimes they will only have access to vendor-supplied finished products that have undergone unknown procedures to make the visual appearance more appealing. Mappers can use shaded relief images derived from LiDAR DEMs or TINs for on-screen digitizing, as well as new derivative products such as terrain slope and LiDAR intensity to identify geological and other features. The new generation of LiDAR data users will be interested in the absolute vertical accuracy of elevations and will need to know how land cover type affects that accuracy.

References

BALTSAVIAS, E. P. 1999. Airborne laser scanning: basic relations and formulas. *ISPRS Journal of Photogrammetry and Remote Sensing*, **54**, 199–214.

QUADE, D. J., GIGLIERANO, J. D. & BETTIS, E. A. III 2004. Surficial geologic materials of the Des Moines Lobe of Iowa, Phase 6: Dickinson and Emmett Counties: IDNR/IGS, OFM-0402, 1:100 000 scale. Available online at: http://www.igsb.uiowa.edu/gsbpubs/pdf/ofm-2004–2.pdf.

WEHR, A. & LOHR, U. 1999. Airborne laser scanning – an introduction and overview. *ISPRS Journal of Photogrammetry and Remote Sensing*, **54**, 68–82.

Monitoring coastal change using terrestrial LiDAR

P. R. N. HOBBS[1]*, A. GIBSON[2], L. JONES[1], C. PENNINGTON[1], G. JENKINS[1],
S. PEARSON[1] & K. FREEBOROUGH[1]

[1]*British Geological Survey, Kingsley Dunham Centre, Keyworth,
Nottingham NG12 5GG, UK*

[2]*School of Earth & Environmental Sciences, University of Portsmouth,
Portsmouth PO1 3QL, UK*

Corresponding author (e-mail: prnh@bgs.ac.uk)

Abstract: The paper describes recent applications by the British Geological Survey (BGS) of the technique of mobile terrestrial Light Detection And Ranging (LiDAR) surveying to monitor various geomorphological changes on English coasts and estuaries. These include cliff recession, landslides and flood defences, and are usually sited at remote locations undergoing dynamic processes with no fixed reference points. Advantages, disadvantages and some practical problems are discussed. The role of GPS in laser scanning is described.

The use and application of terrestrial-based Light Detection And Ranging (LiDAR), using a method popularly known as laser scanning, has greatly increased over the past 5 years. The perception of the technique has changed from that of a novel, but complex, surveying tool to a relatively simple, almost routine, method for precision measurement. The method was first widely used within the quarry industry where the results of repeat surveys were used to manage and plan material extraction. The technique has subsequently found architectural, civil engineering and industrial applications, and, more recently, has been adopted by the computer games industry to capture street scenery. Within geoscience, terrestrial LiDAR has been applied to the monitoring of volcanoes (Hunter *et al.* 2003), earthquake and mining subsidence, quarrying, buildings, heritage and conservation, forensics (Paul & Iwan 2001; Hiatt 2002), landslides (Rowlands *et al.* 2003) and coastal erosion (Hobbs *et al.* 2002; Miller *et al.* 2006; Poulton *et al.* 2006). The method has developed in parallel with airborne LiDAR and, to some extent, with terrestrial photogrammetry (Adams *et al.* 2003) as well as other airborne/spaceborne techniques (Balson *et al.* 1998; Webster & Dias 2006). This paper describes the different techniques and applications to which the British Geological Survey (BGS) has used terrestrial LiDAR over the past 10 years, and the successes and difficulties that have been encountered over that time.

Terrestrial LiDAR: applications at the BGS

Since 1999 the BGS has used various terrestrial LiDAR and GPS systems in combination to measure, record and monitor a variety of geological exposures and geomorphological subjects, initially in collaboration with 3DLaserMapping Ltd. Most of the work has centred on the monitoring of active landslides on eroding coastlines, where the target surface is visible from a number of locations and is generally free of vegetation. Good reflections are returned from natural rock and soil materials at these sites, with rare exceptions where water seepage dramatically reduces the reflectance of dark mudrocks.

The platform for the scanner is usually a tripod (Fig. 1). This provides the versatility and mobility essential when scanning in a dynamic environment, where any kind of permanent installation is ruled out. The instrument can either be positioned over a known point or a differential global positioning system (dGPS) antenna substituted for the scanner to obtain the position, provided that the height is accurately measured, and that the discrepancy between antenna and scanner heights is accounted for. Care must be taken to ensure tripod stability, particularly in sand. Most cliffs can be laser-scanned from the beach or rock platform using this method, but with certain caveats described later.

Perhaps surprisingly, the method is also suited to low-lying features, normally only considered for airborne LiDAR, with the proviso that elevated vantage points are necessary. These may be provided by a vehicle roof in a temporary configuration (Fig. 2) or, increasingly commonly, a dedicated vehicle mounting. Most vehicle-mounting arrangements suffer in strong winds, although jacks can be used at the four corners of the vehicle to eliminate suspension movement and provide stability. A modular gantry (Fig. 3) or hydraulic platform may

From: FLEMING, C., MARSH, S. H. & GILES, J. R. A. (eds) *Elevation Models for Geoscience.*
Geological Society, London, Special Publications, **345**, 117–127.
DOI: 10.1144/SP345.12   0305-8719/10/$15.00 © The Geological Society of London 2010.

Fig. 1. Laser scanner mounted on a tripod: mobile set-up.

also be used. However, there are important issues of stability, particularly where the instrument cannot be remotely operated. In the case of critical monitoring (e.g. for a large extremely hazardous landslide or volcano), a permanent solid monument is preferred and, if possible, the instrument should be mounted on a long-term basis. The latter situation minimizes errors due to instability and setting up

Fig. 2. Laser scanning from a vehicle roof: mobile set-up.

Fig. 3. Laser scanning from a 6 m-high modular gantry: temporary set-up.

associated with temporary tripod-mounting; but, of course, this ties up the instrument for long periods and may expose it to damage. In a coastal situation, the installation of a solid monument is usually not feasible. The set-up shown in Figure 4, whilst providing a solid platform, can only be temporary as the tide covers the block, as evidenced by the mussels attached to it.

Fig. 4. Laser scanner mounted to a rock-bolt on a World War II concrete block: temporary set-up.

In most cases it is not possible to erect permanent monuments on the coast or estuary from which to use the laser scan. Where monitoring is required using temporary mobile platforms (e.g. tripods), laser scans must be oriented to a fixed grid reference system. In areas of coastal erosion lacking fixed reference points, the current solution is geodetic-quality dGPS. The raw output data from a scan consist of vertical and horizontal angles, distances and reflective intensities, plus calibrated digital images where available. The angles and distances are subsequently 'oriented' into *xyz* grid co-ordinate positions (local or global) on the computer using dGPS (or other) location information. This format allows 'oriented' scans taken from other positions to be combined, as well as roving GPS datasets, to form a single model. Recently, the additional feature of calibrated digital photography has enhanced the method, allowing both the raw data and the final three-dimensional (3D) model to be coloured accurately, the outcome resembling a 3D photograph. This is very useful for the geoscientist who wishes to visualize, record and measure terrain, structures, volumes and processes.

Coastal recession

Coastal recession is of worldwide concern, particularly in the light of current global climate change predictions, associated sea-level changes and increased storm occurrence (Clayton 1989; Lee & Clark 2002). Monitoring of recession is considered a key factor in successful coastal management and hazard mitigation (Hall *et al.* 2002). The coastal environment is one in which high-precision surveying can be made difficult by the dynamic nature of the environment. Typically, away from the built environment there are no lasting reference points with which to 'fix' each survey. Each element of the eroding coastline, that is cliff, platform and beach – such as those in many parts of eastern and southern England – is in an almost continuous state of flux. Tides, unstable slopes and the routine destruction of any fixed reference points therefore create an immediate problem for the surveyor using terrestrial LiDAR in these environments. Hence, the need to accurately fix scans based on mobile or temporary set-ups, such as those shown in Figures 1 and 2. The use of dGPS to locate laser scans can itself be compromised close to high cliffs, particularly where the satellite configuration is unfavourable.

Methods other than terrestrial LiDAR have been used to monitor unstable cliffs. These may be subdivided into direct and other remote techniques: examples of the former include instrumented rock-bolts and cable tensiometers; and examples of the latter are time domain reflectometry (Pasuto *et al.* 2000) and digital image processing (Allersma 2000).

As with other sciences, sophisticated computer models are increasingly being used to characterize and predict coastal cliff recession (Walkden & Hall 2005), particularly where the element of climate change is introduced. These require quantitative input data; such as those obtainable by terrestrial and aerial LiDAR, and by other remote techniques. Direct slope stability monitoring methods tend to be targeted at specific features where movement is expected. They, therefore, provide only local information and, crucially, may miss unforeseen movements or events. Quantitative information about mass behaviour usually requires a remote method.

Slope Dynamics Project

The British Geological Survey has been carrying out coastal monitoring in England using terrestrial LiDAR since 1999 (Hobbs *et al.* 2002). The Slope Dynamics Project has 12 coastal locations where 'soft' rock cliffs are subject to marine erosion and/or landslide activity. These sites are scanned annually or bi-annually (depending on the rate of change) to assess the influence of geology, geomorphology and geotechnics on the process of cliff recession. Recently, active inland landslides have been included (Rowlands *et al.* 2003), although these tend to be more unusual and less dynamic than their coastal counterparts. As part of the project, platforms and beaches are included in the scans so that the relationships between wave attack and cliff degradation can be examined. The role of landslides on cliff recession is a topic of some interest in Britain, particularly along the east and south coasts of England where the rocks making up cliffs and platforms are comparatively weak, susceptible to erosion and instability, and marine attack is powerful. Huge quantities of sediment liberated from the cliffs are moved along the coast or offshore, and redeposited by the sea. Modelling this action in response to time and environmental conditions is key to understanding the likely effects of climate change. Such models require quantitative information about cycles of sediment supply, and the relationship with platform erosion and beach thickness, in order to calibrate their predictions. Terrestrial LiDAR is one way of doing this, albeit on a local scale.

The method used by the project in the coastal environment involves setting up a baseline on the foreshore, parallel to the cliff with a tripod at each end (Fig. 5). A laser scanner and a dGPS antenna alternately occupy the tripods. The scanner fixes the position of the other tripod, using either a

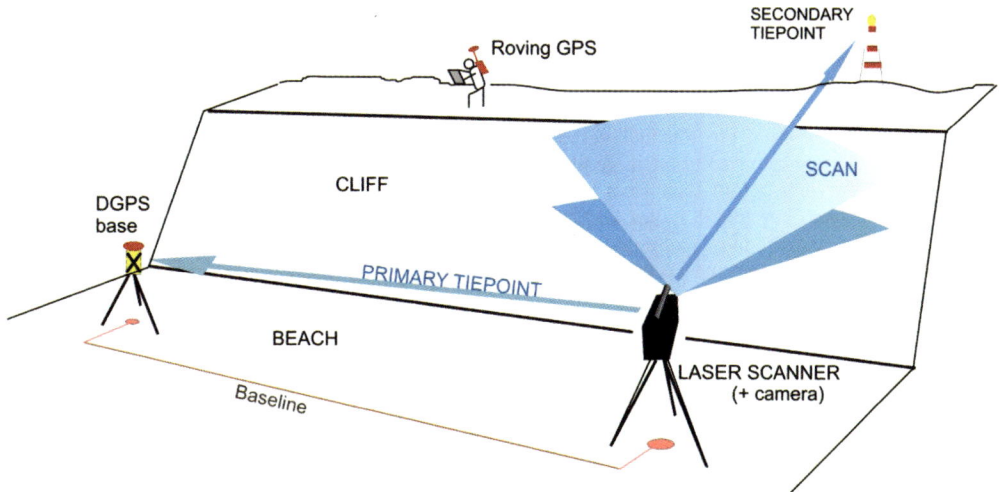

Fig. 5. Coastal mobile terrestrial LiDAR method.

single shot or an automated microscan, to a reflective target in place of the dGPS antenna, and then scans the cliff and platform. This may be repeated along the foreshore or, in some cases, on the cliff top in order to get the fullest possible coverage. Where large-scale rotational landslides and large embayments are present, this task may be difficult, particularly where access to the cliff is impossible or unsafe. Coverage of 'shadow' areas may be improved by infilling with a roving GPS where access is possible. The scan data consist of *xyz* position and reflective intensity, with the possible addition of a digital image mosaic provided by a built-in calibrated camera. The final output can be in the form of a 'point-cloud' (Fig. 6), a 2D intensity plot (Fig. 7), a 3D point-cloud coloured from the

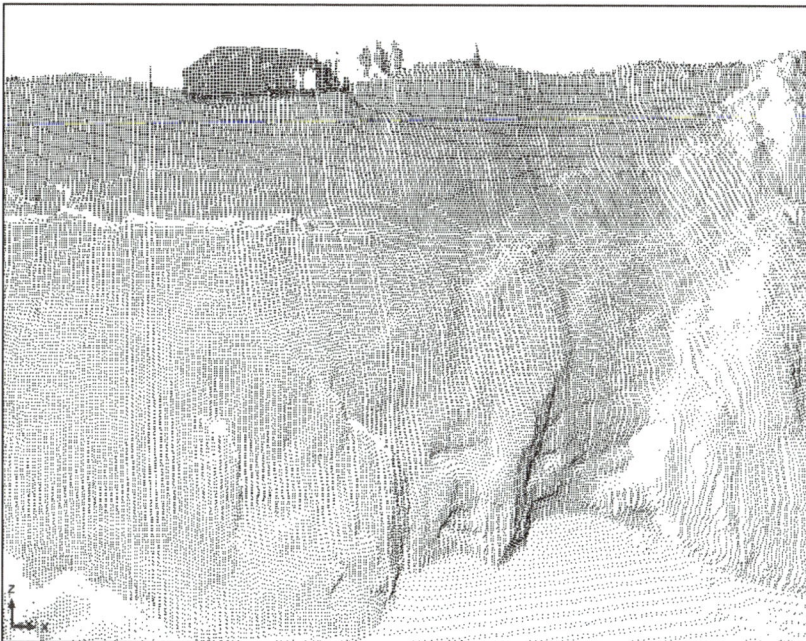

Fig. 6. Part of Scan A showing a raw point-cloud.

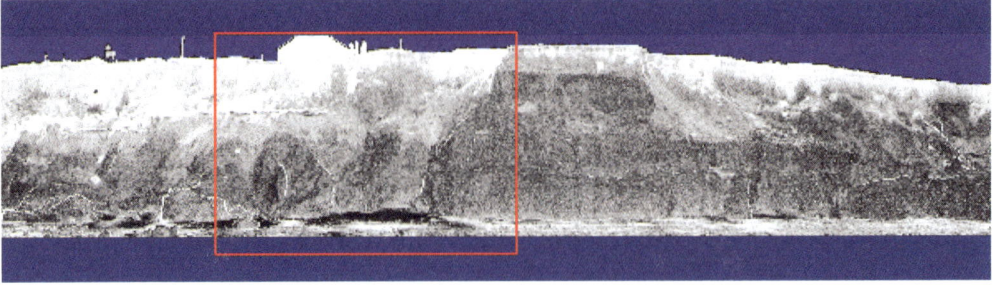

Fig. 7. Full Scan A: 2D intensity plot.

photo-mosaic (Fig. 8) or a 3D triangulated 'solid' surface model (Fig. 9) over which the photo-mosaic has been draped (Fig. 10). The coloured point-cloud output (Fig. 8) is visually effective where the density of points is high, but the solid model allows greater manipulation and the calculation of areas, volumes and cross-sections. False-colour models can be utilized to show height (Fig. 11) and range (Fig. 12). The various uses of these models by the geoscientist are summarized in Table 1.

The 'Scan A' example shown in Figures 6–13 is a 20 m-high cliff formed in matrix-dominant Late Devensian tills (Withernsea Till and Skipsea Till members of the Holderness Formation), from part of the 50 km-long Holderness coast of East Yorkshire. Historical erosion rates are between 1 and 2 m year^{-1} (Balson *et al.* 1998). Landslides on this coastline typically consist of single rotational failures and smaller toppling failures. The rotational features tend to develop en echelon, a factor possibly related to subvertical joint patterns in the tills.

Fig. 8. Full Scan A: 3D coloured point-cloud.

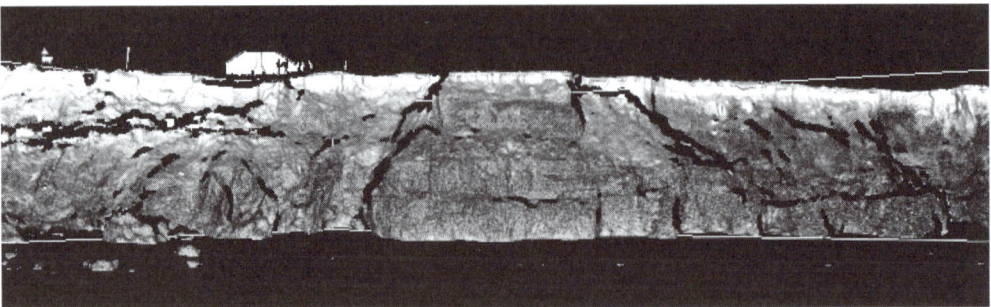

Fig. 9. Full Scan A: 3D triangulated 'solid' surface model.

Fig. 10. Full Scan A: 3D triangulated surface model with a digital colour photograph overlay.

It is clear from the figures depicting Scan A that each model contains gaps or 'shadow' areas, which represent areas that the laser was unable to 'see'. For instance, in Figure 13 a boulder close to the scanner has cast a long laser 'shadow' across the beach thus preventing any points being captured behind it. This can be rectified to some extent by carrying out multiple scans from several positions, each having a different viewpoint on the same object. Then with the application of the dGPS, or other 3D model orientation method, these 'shadows' can be significantly reduced or eliminated. Of course, this adds

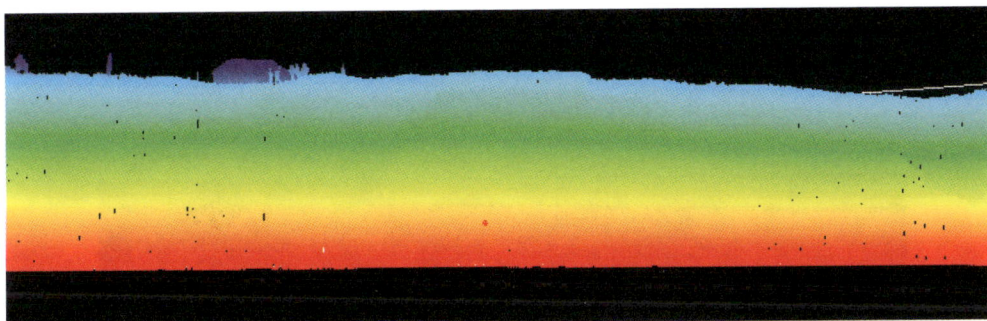

Fig. 11. False-colour 2D height model for Scan A (red, low; blue, high).

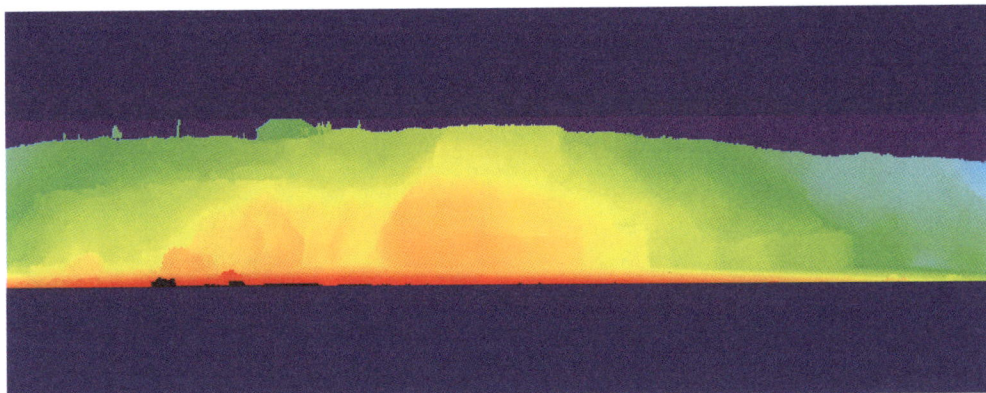

Fig. 12. False-colour 2D range model for Scan A (red, near; blue, far).

Table 1. *Uses for each model type*

Uses	Model				
	2D intensity	3D point-cloud	3D colour point-cloud	3D solid	False-colour
Lithology recognition	✓		✓	✓	
Geomorphology		✓	✓	✓	✓
Structural geology			✓	✓	
Volumes, areas				✓	✓
Cross-sections				✓	

considerably to the amount of post-processing required. New dGPS systems are reducing the amount of post-processing by improved real-time processing; for example, by mobile telephone communication with a GPS network. However, this may introduce a fresh problem associated with mobile network coverage in remote locations.

Problems to consider

One problem with the triangulated 'solid' surface is that uneven coverage of points in the original point-cloud results in either gaps in the model (Fig. 9) or oversized polygons (Fig. 14) depending on the threshold parameters selected. This is particularly the case where the cliff is receding unevenly from crest to toe (i.e. it has a 'stepped' profile) or where

landslides are of a rotational type, featuring back-tilted slip masses, and hence are in the laser 'shadow' when scanned from the beach. As the laser scanner sweeps the subject from its fixed position it has the attribute of a shotgun; that is, nearby features are densely covered with points compared with distant features. In the case of a largely planar subject such as a building, this may not be a problem. However, for natural features such as cliffs, the result may be wide variation in the surface detail captured, and hence the integrity of the 3D solid model.

In the case of large coastal landslide complexes, the range of the instrument becomes an issue. Many high-speed laser scanners, with a maximum range of typically less than 500 m, might struggle with such features, particularly where access to the cliff to carry out multiple scans is impossible. A common

Fig. 13. Side view of a part of Scan A: 3D raw point-cloud (red arrow, boulder casting a shadow).

Fig. 14. Scan B: 3D triangulated surface model.

problem encountered during the project has been the inadequacy of PC–software combinations to deal with the millions of points produced by modern scanners, notwithstanding that the scanners used were not classed as 'high-speed'. This is particularly the case where 'solid' models are required. The repeated scanning of the same cliff enables changes in elevation to be displayed and quantified,

provided that a solid grid-oriented model for each epoch has been produced.

Change models

The 'Scan B' example shown in Figures 14 and 15 is of a cliff up to 50 m high on the North Yorkshire

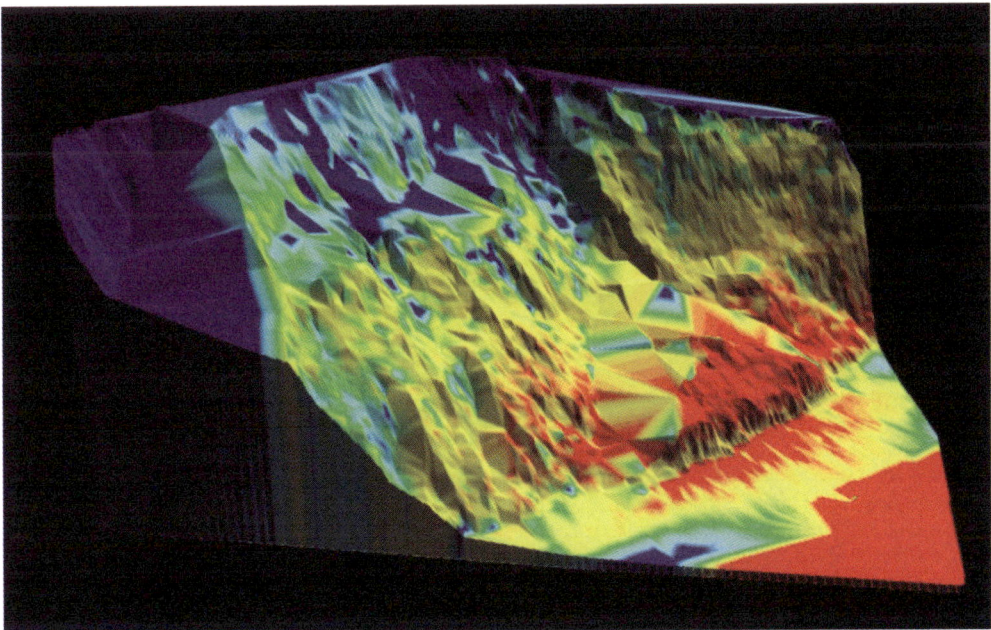

Fig. 15. Scan B: 3D elevation 'change' model for part of Scan B (refer to Fig. 14) (red, height increase; blue, height decrease).

coast consisting of complex superficial deposits of till and other interlayered glacial deposits overlying folded and weak Speeton Clay Formation mudrocks. The 'change' model in Figure 15 shows the elevation difference between two solid 3D models, derived from scans taken 1 year apart. The resulting annual vertical displacement is coloured proportionately so that red is maximum ground-level rise and blue is maximum ground-level fall, although the two are not of equal magnitude. In terms of slope morphology, the change model shows us that a debris flow at the toe of the cliff has accumulated material, the backscarp has lost material and changes have occurred to the beach levels. Information in the area of the oversized polygons to the rear of the debris flow is probably unreliable. Again, the density and reliability of data are important factors when interpreting these change models (Miller *et al.* 2006).

When considering change models of coastal cliffs it is important to include the foreshore (platform and beach) as part of the same model as the cliff itself. This is for two reasons: first, the beach is a transient feature that may consist of both transported material and debris derived locally from the landslide. As such, support of the cliff toe and restriction of seepage are also transient features affecting the cliff itself. Secondly, a deep-seated landslide may have its slip surface extending below the level of the foreshore and, hence, the foreshore becomes directly involved in the movement. In the long term, the erosion of the platform itself must be considered as part of the model (Walkden & Hall 2005).

The methodology for producing 'change' models is very much dependent on the software packages available to the user, and can be achieved in a variety of ways and with variable effectiveness depending on the geometry of the subject. These usually involve more than one package, and possibly as many as four. Issues relating to the robustness of such models have been addressed by Miller *et al.* (2008).

Conclusions

In order to correctly interpret the terrestrial LiDAR change models described, a combination of models should be considered. These could include the elevation and range models (Figs 11 & 12), which can have their own change models derived so that the vertical and horizontal components of movement can be resolved. In its simplest form, a rock fall from a vertical cliff face represents a loss of material from the cliff face and an accretion of debris on the beach. However, a similar fall from the crest of an inclined cliff will result in accretion at mid-cliff. This might appear from an elevation change model alone as if caused by an uplift of strata – as, for example, at the toe of a rotational landslide – rather than a deposition of fallen debris. Subtle precursor processes such as the opening of joints may produce apparent 'accretion' of the cliff face prior to failure and ultimate recession. Such small movements may or may not be resolved by the scan depending on the method and equipment used.

The basic 2D intensity model (Fig. 7) is useful in distinguishing textures. This has a greater applicability to man-made structures and materials (e.g. concrete, metal, brick), but can still be useful for distinguishing the lithologies of strongly contrasting geomaterials. Laser-scan models enhanced by calibrated digital photography (Figs 8 & 10) are a useful resource for the geoscientist, as the textures and colours greatly enhance the interpretation of lithology, structure and geomorphology. This is, of course, further enhanced by the 3D capability, whereby the true geometry of coastal landslide and erosion features can be appreciated. Solid 3D models allow volumes and areas to be calculated either in relation to a planar datum or to a previous model. Thus, displaced volumes may be calculated and displaced masses estimated.

References

ADAMS, J. C., SMITH, M. J. & BINGLEY, R. M. 2003. Development and integration of terrestrial cliff-face mapping techniques with regional coastal monitoring. *In*: APLIN, P. & MATHER, P. M. (eds) *Proceedings of RSPSoc 2003: Scales and Dynamics in Observing the Environment, Nottingham, England.* Remote Sensing and Photogrammetry Society, Nottingham.

ALLERSMA, H. G. B. 2000. Measuring of slope surface displacement by using digital image processing. *In*: BROMHEAD, E., DIXON, N. & IBSEN, M.-L. (eds) *Landslides in Research, Theory, and Practice.* Thomas Telford, London, **1**, 37–44.

BALSON, P. S., TRAGHEIM, D. & NEWSHAM, R. 1998. Determination and prediction of sediment yields from recession of the Holderness coast, Eastern England. *In*: *Proceedings of the 33rd MAFF Conference of River and Coastal Engineers, 1998.* Ministry of Agriculture, Fisheries and Food, London, 4.5.1–4.6.2.

CLAYTON, K. M. 1989. Sediment input from the Norfolk cliffs, Eastern England – a century of coastal protection and its effects. *Journal of Coastal Research*, **5**, 433–442.

HALL, J. W., MEADOWCROFT, I. C., LEE, E. M. & VAN GELDER, P. H. A. J. M. 2002. Stochastic simulation of episodic soft coastal cliff recession. *Coastal Engineering*, **46**, 159–174.

HOBBS, P. R. N., HUMPHREYS, B. *ET AL.* 2002. Monitoring the role of landslides in 'soft cliff' coastal recession. *In*: MCINNES, R. G. & JAKEWAYS, J. (eds) *Instability: Planning and Management.* Thomas Telford, London, 589–600.

HIATT, M. E. 2002. Sensor integration aids mapping at ground zero. *Photogrammetric Engineering and Remote Sensing*, **68**, 877–879.

HUNTER, G., PINKERTON, H., AIREY, R. & CALVARI, S. 2003. The application of a long-range laser scanner for monitoring volcanic activity on Mount Etna. *Journal of Volcanology and Geothermal Research*, **123**, 203–210.

LEE, E. M. & CLARK, A. R. 2002. *Investigation and Management of Soft Rock Cliffs*. Thomas Telford, London.

MILLER, P. E., MILLS, J. P., EDWARDS, S. J., BRYAN, P., MARSH, S. & HOBBS, P. 2006. Integrated remote monitoring for coastal geohazards and heritage sites. *In*: *Proceedings of the ASPRS 2006 Annual Conference, Reno, Nevada, May 1–6 2006*. On CD Rom. American Society for Photogrammetry & Remote Sensing, Bethesda, MD.

MILLER, P. E., MILLS, J. P., EDWARDS, S. J., BRYAN, P. G., MARSH, S. H., HOBBS, P. & MITCHELL, H. 2008. A robust surface matching technique for coastal geohazard monitoring. *ISPRS Journal of Photogrammetry and Remote Sensing*, **63**, 529–542.

PASUTO, A., SILVANO, S. & BERLASSO, G. 2000. Application of time-domain reflectometry (TDR) technique in monitoring the Pramollo Pass landslide (Province of Udine, Italy). *In*: BROMHEAD, E., DIXON, N. & IBSEN, M.-L. (eds) *Landslides in Research, Theory, and Practice*. Thomas Telford, London, **3**, 1189–1194.

PAUL, F. & IWAN, P. 2001. Data collection at major incident scenes using three dimensional laser scanning techniques. *In*: *The Institute of Traffic Accident Investigators: 5th International Conference held at York*. The Institute of Traffic Accident Investigators, Shrewsbury.

POULTON, C. V. L., LEE, J. R., HOBBS, P. R. N., JONES, L. & HALL, M. 2006. Preliminary investigation into monitoring coastal erosion using terrestrial laser scanning: case study at Happisburgh, Norfolk. *Bulletin of Geological Society of Norfolk*, **56**, 45–64.

ROWLANDS, K., JONES, L. & WHITWORTH, M. 2003. Photographic feature: landslide laser scanning: a new look at an old problem. *Quarterly Journal of Engineering Geology*, **36**, 155–158.

WALKDEN, M. J. & HALL, J. W. 2005. A predictive mesoscale model of the erosion and profile development of soft rock shores. *Coastal Engineering*, **52**, 535–563.

WEBSTER, T. L. & DIAS, G. 2006. An automated GIS procedure for comparing GPS and proximal LIDAR elevations. *Computers & Geoscience*, **32**, 713–726.

Digital elevation models for research: UK datasets, copyright and derived products

M. J. SMITH

School of Geography, Geology and the Environment, Kingston University,
Kingston-upon-Thames, Surrey KT1 2EE, UK (e-mail: michael.smith@kingston.ac.uk)

Abstract: The UK is served by a wide range of digital elevation models (DEMs) that have a variety of technical specifications from several different vendors. The abundance of data presents researchers with a complex range of choices dependent upon their application (and therefore 'fitness-for-purpose') and desired use of intellectual property rights (IPR). This paper explores current DEM datasets of the UK and presents their use within the context of claimed copyright and IPR. In particular, responsibilities placed upon grant holders for the lodgement of research outputs by UK Research Councils places new emphasis upon data access, derived data and data re-use. The complex interplay of rights between research output stakeholders (data suppliers, data creators, data users) presents a difficult scenario for both data repositories and data depositors.

Digital elevation models (DEMs) form important input parameters for a variety of environmental modelling projects, ranging from intervisibility analyses, through to flood modelling and tree-canopy height estimation (e.g. Lane *et al.* 2001). DEMs are therefore integral to a range of qualitative and quantitative environmental methodologies; Smith & Pain (2009) provide a review of the use of remote sensing, including DEMs, in geomorphology. Data sources for the creation of DEMs have traditionally been contours that were photogrammetrically derived (e.g. Evans 1972); however, these have more recently been supplemented with data from techniques such as digital photogrammetry (e.g. Chandler 1999), interferometric synthetic aperture radar (or IfSAR) (e.g. Muller *et al.* 2000) and Light Detection And Ranging (or LiDAR) (e.g. Arnold 2006). All of these techniques are available, to some extent, from both airborne and satellite-borne platforms. For example, photogrammetry can be performed using aerial photography or satellite imagery, whilst DEMs have been generated using spaceborne (e.g. Shuttle Radar Topography Mission or SRTM) and airborne (e.g. Intermap's NEXTMap product) IfSAR. The wide range of data (often from different vendors) derived from different techniques, using different platforms and at different dates, produces a complex 'map' of available products.

The UK is one of the most mapped nations on Earth, with the Ordnance Survey (OS) at the forefront of high spatial and temporal resolution data acquisition (e.g. Mastermap provides object accuracy of *c.* 1 m and currency of *c.* 6 months: Ordnance Survey 2006). With the increasing use of digital elevation data, data suppliers have made available at least eight different DEM products

(including global datasets) using a variety of data acquisition techniques and, consequently, with varying technical specifications. In addition to data technical specifications, vendors may require the use of a licence agreement to restrict how data may be used. In particular, this will involve publication restrictions and intellectual property rights (IPR) claims over derivative datasets. The following sections outline the main DEM datasets available for the UK and places them within the context of data licensing; the latter is particularly important with respect to derivative data products where academics may wish to distribute and re-use research outputs.

Digital elevation models

Digital elevation data have traditionally been collected using ground survey; however, it is a slow and relatively labour-intensive technique. The advent of photogrammetry allowed the rapid production of contours and this has been the primary method for the first generation of national DEM data products. More recent developments have seen the ability to digitally capture and process data over large areas using digital photogrammetry, IfSAR and LiDAR. These technological drivers, coupled with the extensive exploitation of geographical data by government bodies and companies, have made the UK a focus for the provision of DEM data. Subsequent to data collection, final products are usually interpolated to a regular raster grid for distribution. During data collection, IfSAR and LiDAR are based upon sampling as close to a regular grid as possible; both ground survey and digital photogrammetry approach this set-up, although may deviate from a grid collection procedure dependent upon implementation.

From: FLEMING, C., MARSH, S. H. & GILES, J. R. A. (eds) *Elevation Models for Geoscience.*
Geological Society, London, Special Publications, **345**, 129–133.
DOI: 10.1144/SP345.13   0305-8719/10/$15.00 © The Geological Society of London 2010.

Interpolation to a regular grid is therefore a straight-forward procedure. As contours are isolines, joining points of equal elevation, the contour interval influences the density of measurements. However, density is also controlled by slope angle, with steeper slopes generating a greater density of measurements. Contours, therefore, have a variable measurement density with fewer measurements in areas of low slope. Consequently, DEMs derived from contours are more complex to interpret within a modelling situation.

This section briefly describes the main national DEM datasets available for the UK; summary information is displayed in Table 1, including individual dataset names, nominal resolution, vertical accuracy and acquisition date. Whilst not listed here, space-borne sensors are routinely used to collect stereo-scopic data, subsequently being processed into DEMs (e.g. SPOT5, ASTER); however, coverage is variable. There are broadly four main providers of national DEMs in the UK.

(1) *Ordnance Survey*: the OS has two main DEM products that are available to the higher education (HE) community under the JISC–OS licence (distributed through Digimap). These are named Land-Form Panorama® and Land-Form Profile®, and are based on 1:50 000 and 1:10 000 topographical contours, respectively. A third product, called Land-Form ProfilePlus®, is a composite dataset based on contours, digital photogrammetry and LiDAR, and is available

commercially (Ordnance Survey 2005). Both Panorama® and Profile® raster DEMs are interpolated from the original contours.

(2) *Intermap*: the NEXTMap DEM is based on airborne IfSAR, allowing significantly better modelling of the terrain surface than any other current national product (Smith *et al.* 2006). It primarily competes with LiDAR, offering economic survey costs; however, it is unable to penetrate vegetation and, in general, has lower vertical accuracies (Mercer 2001).

(3) *Joint Information Systems Committee (JISC)*: the Landmap DEM was created by a JISC-funded project (Muller *et al.* 2000) using spaceborne IfSAR and is free from third-party licensing. Whilst IfSAR typically utilizes two SAR sensors on a single platform ('single-pass'), Landmap used data from two temporally separated sensors (called ERS-1/2 tandem data). In general, DEMs generated using tandem data are of lower quality than comparable single-pass IfSAR (Rosen *et al.* 2000). Uniquely, the DEM was produced using strip data rather than individual scenes.

(4) National Aeronautics and Space Administration (NASA)–Deutsches Zentrum für Luft- und Raumfahrt (DLR)–Agenzia Spaziale Italiana (ASI): SRTM (Rabus *et al.* 2003), a NASA Space Shuttle mission, created a DEM of the world from 56°S to 60°N. SRTM employed single-pass spaceborne IfSAR and, uniquely, carried two sets

Table 1. *National DEM data products of the UK*

	Creator	Nominal resolution (m)	Relative vertical accuracy (m)	Acquisition technique	Acquisition date
Panorama®	OS	50	5	Photogrammetry (contours)	Maintained until 2002
Profile®	OS	10	5	Photogrammetry (contours)	Maintained
ProfilePlus®	OS	2–10	0.5–2.5	Multiple	Maintained
NEXTMap	Intermap	5 10	Type 1: 0.5 Type 2: 1	InSAR (airborne)	2002–2003
Landmap	JISC	25	20	Tandem InSAR (spaceborne)	1995–1996
SRTM:					
C-band 30 m	NASA	30	6	IfSAR (spaceborne)	March 2003
C-band 90 m	NASA	90	6		
X-band	DLR–ASI	25	6		
LiDAR (limited availability)	Various	0.5–5	0.10–0.25	LiDAR	Variable

Note: LiDAR is included for comparison; whilst not available nationally, large parts of the country have had data collected for them.

of sensors on board. These belonged to NASA (C-band) and DLR–ASI (X-band), and have been processed/distributed separately. The latter was able to operate at a higher level of accuracy but only covered approximately 50% of the area of the C-band sensor.

Copyright: third party and derived data

When using third party geospatial data under licence, the vendor will typically claim copyright to the original dataset. The licence will then restrict the manner in which the data can be used, copied or redistributed in order to protect the intellectual property of the vendor. Such licences are clearly pertinent to the use of DEMs as users may wish to publish the results of their research. Two main issues subsequently pertain to research outputs.

(1) *Publication of results*: researchers may want to publish research results (non-commercially) in academic journals. Licences will often restrict the graphical publication of data based upon the publication medium, graphic size and area of study.

(2) *Derived data*: DEMs are often used within environmental modelling (e.g. flood modelling), where the primary research outputs are derived from several input datasets (including DEMs). The vendors of each of the input datasets are potentially stakeholders in the output dataset and may claim IPR. Such a claim would subsequently effect any publication or distribution of the output dataset.

Whilst there is currently much debate concerning access to publicly collected data in the UK (e.g. Arthur & Cross 2006), this is largely irrelevant to the issues discussed in this paper. Even though users would clearly prefer freely accessible data (notwithstanding arguments of data quality), it is the licence agreement itself, and restrictions based around publishing and redistribution, that are pertinent. However, it should be noted that this debate is taking place against the backdrop of freely accessible, federally collected, data in the United States that is typically free of intellectual property rights claims (e.g. SRTM, Landsat), distributed at the of cost reproduction and allows unrestricted use within other products. The following outlines the issues of publication and derived data with respect to each of the data vendors.

(1) *Ordnance Survey*: the OS has historically been the predominant supplier of DEM data in the UK; both Panorama® and Profile® are extensively used within universities. The use of OS data in posters and presentations is relatively open and generous; however, all electronic, Internet-facing documents come with very stringent restrictions (EDINA 2006) based on the size of each individual image and the ground area covered. Specific details are based on the individual product, but generally means that any map larger than A5 is unpublishable. If a secondary data product is (partially) derived from an OS data source (e.g. Panorama®) then the OS claims IPR over that product, meaning that it inherits the same IPR restrictions as the source data. Specifically, this means no data redistribution and publication restrictions. It should be noted that once a licence has lapsed the source data and all derived data must be destroyed. Dornbush *et al.* (2006) used OS data to georectify historic mapping and, whilst the OS claims IPR to the derived product (a georectified historic image), they have allowed publication of the final product. Conversely, Lukas & Lukas (2006) extensively modified OS contour data as part of a basemap, but were refused permission to publish their map. Publication was achieved within the licence restrictions, but required the production of 12 map panels split across two articles.

(2) *Intermap*: Intermap do not allow the redistribution of their data by end-users; however, they include a 'thematic data' clause in their licence (Intermap Technologies 2005) that allows the end-user to publish and redistribute derived data products. Specifically, the clause requires that the final output cannot be reverse-engineered to the original source data. In such an instance Intermap do not claim IPR and, therefore, the final product is free from the inheritance of IPR.

(3) *JISC*: JISC licence the Landmap product under copyright (Landmap 2006) and, as with other vendors, restrict its reproduction and distribution. However, the licence explicitly allows images to be published within research work and does not claim IPR over derivative products.

(4) *NASA*: the C-band SRTM data have been released under a 'memorandum of understanding' between NASA and the National Geospatial-Intelligence Agency (NGA) that allows the data 'to be archived in the appropriate NASA data center and made available to all users without restriction and at no more than the cost of fulfilling the user request' (USGS 2006). As such the data are 'in the public domain'.

(5) *DLR–ASI*: the X-band SRTM data is copyright to the DLR–ASI and distributed by the DLR. The licence (DLR 2006) restricts data

reproduction and distribution, whilst implicitly claiming IPR over derivative products. Publication of research outputs is unrestricted.

Discussion

The UK has an unusually rich and diverse set of national digital elevation data products available for use. The OS products have been widely used within universities for both research and teaching under the JISC–OS licence; however, the development of more recent products has led to greater choice. Broad technical specifications highlight that LiDAR is the elevation data product of choice; however, lack of national coverage, cost and data volume means that it may not be suitable for individual users. IfSAR-based NEXTMap is currently the highest specification national product currently available, and is seeing wide use in both the commercial and academic sectors. For example, Norwich Union, the British Geological Survey and the Environment Agency all licence the entire dataset. Whilst both Panorama® and Profile® (from the OS) compete with NEXTMap, they are generally considered to be inferior (e.g. Smith et al. 2006). ProfilePlus® is intended to fill this gap; however, as a composite product, consistency is spatially variable and, therefore, care must be taken in its use. Landmap and SRTM are satellite-based IfSAR products and, consequently, their technical specifications are substantially lower than products from either Intermap or the OS.

Unfortunately, technical specifications are now often considered of secondary importance to IPR-related issues. The ability to publish research results, and retain IPR over derived products, is of paramount importance to the commercial re-use of outputs partially derived from elevation data. Within the university sector the government has highlighted the need to access research results (HMSO 2004), whilst Research Councils UK now requires the deposition of research outputs from grant holders (Research Councils UK 2005), including both publications and data. In addition to NEXTMap being technically superior to OS national elevation data products, the ability for users to retain IPR over thematically derived outputs makes it an ideal product for both commercial users and researchers. It also raises the potential scenario of one government agency specifically not using OS data (another government agency) solely because of licensing issues.

The area of derivative data products requires careful consideration when one input dataset is used. It becomes complex when a variety of input datasets are combined, such as in flood modelling,

to produce a variety of data outputs. The EDINA GRADE Project (www.edina.ac.uk/projects/grade) is currently researching the legal, social and technical aspects of establishing a geospatial data repository. Within universities, this is a key element in meeting the Research Councils UK requirement for data lodgement and, until issues related to IPR are resolved, such repositories will remain unworkable. The legal (IPR) related part of the project has produced a compendium of examples in the use of geospatial data (Smith 2006) and this is being used as a basis to explore the legal framework (e.g. copyright and database right) that operates between stakeholders in geospatial data output.

Conclusions

This paper has briefly introduced how DEMs are used in academic research, with a particular focus on national UK coverage. There are currently eight national datasets, primarily based on photogrammetrically derived contours and IfSAR; although it should be noted that, whilst LiDAR is often the preferred data collection method, there is currently limited coverage. The range of datasets discussed have been collected using different techniques, from different 'platforms' and at different dates. They therefore represent a complex range of choices for the end-user, with selection dependent on the application and 'fitness-for-purpose'.

Data copyright is an important aspect that should be considered during the selection procedure, in addition to the technical specifications. Commercial products will require licensing that may restrict their use. Within the context of the UK, the JISC–OS licence restricts publication of research ouputs based on original OS data (and derived products) such that research outputs are effectively not publishable in any journal. The OS additionally claims IPR over derived products. In comparison, Intermap claims no IPR over 'thematically derived' data products. Data collected by the US federal government (and released under the Freedom of Information Act) pass into the public domain.

Derived data products are a fundamental aspect of enquiry-driven academic research, and are central to the use, distribution and re-use of research. The importance of the 'results of research' are now formally recognized by Research Councils UK, such that all 'outputs' (publications and data) are required to be lodged with an appropriate data centre. The rights of stakeholders in the IPR of derived data is a central issue in all academic research and it is, therefore, important that researchers consider the implications of commercial licences when selecting input datasets for environmental modelling.

I would like to thank Charlotte Waelde for comments on the manuscript. Since this manuscript was accepted for publication, some OS data has been made freely available for public use, including Panorama DEM (http://data.gov.uk/data/publicbody/ordnancesurvey). Additionally a single contiguous global DEM has been compiled using ASTER data (http://www.gdem.aster.ersdac.or.jp).

References

ARNOLD, N. S. 2006. Evaluating the potential of high-resolution airborne lidar data in glaciology. *International Journal of Remote Sensing*, **27**, 1233–1251.

ARTHUR, C. & CROSS, M. 2006. Give us back our crown jewels. *The Guardian, Technology*, 9 March, 1.

CHANDLER, J. H. 1999. Effective application of automated digital photogrammetry for geomorphological research. *Earth Surface Processes and Landforms*, **24**, 51–63.

DORNBUSCH, U., ROBINSON, D. A., MOSES, C., WILLIAMS, R. & COSTA, S. 2006. Chalk cliff retreat in East Sussex and Kent 1870s to 2001. *Journal of Maps*, **v2006**, 71–78.

DLR 2006. Terms and conditions regarding delivery and use of products from the 'Shuttle Radar Topography Mission (SRTM)'. Available online at: http://www.dlr.de/srtm/produkte/lizenzbestimmungen_en.htm (accessed 31 July 2006).

EDINA 2006. Digimap: Ordnance Survey Data Sublicence Agreement. Available online at: http://edina.ac.uk/digimap/osterms.html (accessed 31 July 2006).

EVANS, I. S. 1972. General geomorphometry, derivatives of altitude and descriptive statistics. *In*: CHORLEY, R. J. (ed.) *Spatial Analysis in Geomorphology*. Harper & Row, New York, 17–90.

HMSO 2004. *Scientific Publications: Free for All? Science and Technology Committee, Tenth Report of Session 2003–04*. Stationery Office, London.

INTERMAP TECHNOLOGIES 2005. Product Handbook. Available online at: http://www.intermap.com/images/handbook/product-handbook.pdf (accessed 21 October 2005).

LANDMAP 2006. Terms for the Licensing of Data for Use by Institutions in the Academic Community in the British Isles and Research Councils in the United Kingdom. Available online at: http://landmap.ac.uk/Registration/landmap_license.htm.

LANE, S. N., CHANDLER, J. H. & PORFIRI, K. 2001. Monitoring river channel and flume surfaces with digital photogrammetry. *ASCE Journal of Hydraulic Engineering*, **127**, 871–877.

LUKAS, S. & LUKAS, T. 2006. A glacial geological and geomorphological map of the far NW Highlands, Scotland, Part 1. *Journal of Maps*, **2006**, 43–55, doi: 10.4113/jom.2006.50.

MERCER, B. 2001. Comparing LiDAR and IfSAR: what can you expect? *In*: FRITSCH, D. & SPILLER, R. (eds) *Proceedings of Photogrammetric Week 2001*. Wichmann, Heidelberg, 227–237.

MULLER, J. P., MORLEY, J. G., ET AL. 2000. The LANDMAP project for the automated creation and validation of multi-resolution orthorectified satellite image products and a 1″ DEM of the British Isles from ERS tandem SAR interferometry. *In*: *RSS2000: Adding Value to Remotely Sensed Data. Proceedings of the 26th Annual Conference of the Remote Sensing Society 12–14th September 2000, University of Leicester*. Remote Sensing Society, Reading.

ORDNANCE SURVEY 2005. Land-Form PROFILE® Plus. Available online at: http://www.ordnance-survey.co.uk/oswebsite/products/landformprofileplus/ (accessed 31 July 2006).

ORDNANCE SURVEY 2006. *Revision Policy for Basic-scale Products*. Ordnance Survey, Southampton.

RABUS, B., EINEDER, M., ROTH, A. & BAMLER, R. 2003. The shuttle radar topography mission – a new class of digital elevation models acquired by spaceborne radar. *Journal of Photogrammetry and Remote Sensing*, **57**, 241–262.

RESEARCH COUNCILS UK 2005. *RCUK Position Statement on Access to Research Outputs*. Research Councils UK Report.

ROSEN, P. A., HENSLEY, S., JOUGHIN, I. R., LI, F. K., MADSEN, S. N., RODRIGUEZ, E. & GOLDSTEIN, R. M. 2000. Synthetic aperture radar interferometry. *Proceedings of the IEEE*, **88**, 333–382.

SMITH, M. J. 2006. *Use Case Compendium of Derived Geospatial Data*. Final report to the GRADE Project. Available online at: http://edina.ac.uk/projects/grade/usecasecompendium.pdf.

SMITH, M. J., ROSE, J. & BOOTH, S. 2006. Geomorphological mapping of glacial landforms from remotely sensed data: an evaluation of the principal data sources and an assessment of their quality. *Geomorphology*, **76**, 148–165.

SMITH, M. J. & PAIN, C. 2009. Applications of remote sensing in geomorphology. *Progress in Physical Geography*, **33**, 568–582.

USGS 2006. Shuttle Radar Topography Missions: USGS's Role. Available online at: http://srtm.usgs.gov/usgsrole.html (accessed 31 July 2006).

Dataset acquisition to support geoscience

J. R. A. GILES, S. H. MARSH* & B. NAPIER

British Geological Survey, Kingsley Dunham Centre, Keyworth, Nottingham NG12 5GG, UK

Corresponding author (e-mail: shm@bgs.ac.uk)

Abstract: Environmental scientists are both producers and consumers of data. Numerous studies have shown that significant amounts of scientists' time can be consumed in acquiring, managing and transforming data prior to their use. To facilitate the work of its scientists, the British Geological Survey (BGS) has identified a series of national datasets that are required by scientists across the organization. The BGS then seeks to acquire and manage these centrally, and to supply them to the scientists in formats that they normally use. Making these datasets readily available helps to:

- enhance the quality of the science;
- promote interdisciplinary working;
- reduce costs.

The strategy has also enabled the development of advanced, domain-specific visualization tools, which have significantly improved the scientific output while also reducing costs.

A modern geological survey organization (GSO), such as the British Geological Survey (BGS), requires a wide range of digital and analogue datasets to support the activities of the scientists employed to fulfil its mission. For many years GSOs primarily used datasets that they had compiled internally. Typically, databases would be designed in-house to meet the needs of specific geoscience communities within the GSO. An example of this is provided by geochemists who analyse and database stream sediments for a specified suite of chemical elements, and visualize the data spatially to understand the distribution of those elements (Johnson *et al.* 2005). In addition to in-house datasets, the legislation in some countries provides GSOs with ready access to specified datasets produced by industry. For example, Geoscience Australia (www.ga.gov.au) houses one of the world's largest collections of petroleum data in its Petroleum Data Repository. This is accessible internally within the GSO, and much of the dataset is 'Open file' and available on the Internet through The Petroleum Information Management System (www.ga.gov.au/oracle/npd/).

During the past decade a range of public and private sector organizations have been creating national digital datasets to meet the needs of a variety of customers. Some of these datasets have direct relevance to geoscience and can be readily integrated with the in-house digital datasets that GSOs typically maintain. For example, Intermap™ Technologies (www.intermap.com) have created a range of regional and national datasets using airborne Interferometric Synthetic Aperture Radar (IfSAR). Under the brand name NEXTMap®, Intermap has produced digital surface models and digital terrain models for Britain, large parts of Europe and several states in the USA.

The process that the BGS undertook to identify, acquire and deliver specific digital datasets to support the work of its geoscientists is described below.

Datasets

In 1999 the BGS geoIDS (BGS Geoscience Integrated Database System) and the SIGMA (System for Integrated Geological Mapping) projects within the BGS established a team to identify the national third-party digital spatial datasets that the BGS was likely to need for the development of modern field capture and 3D modelling activities during the next decade. The team was also tasked to identify potential suppliers and put in place activities to acquire the datasets under appropriate terms and conditions. The team identified the following digital datasets.

- Elevation:
  - terrain and elevation models;
  - topographical survey elevation.
- Imagery:
  - aerial images;
  - satellite images.
- Topography:
  - modern topography;
  - historical topography of various ages.

From: FLEMING, C., MARSH, S. H. & GILES, J. R. A. (eds) *Elevation Models for Geoscience.*
Geological Society, London, Special Publications, **345**, 135–143.
DOI: 10.1144/SP345.14   0305-8719/10/$15.00 © The Geological Society of London 2010.

Elevation

Elevation data are a fundamental tool in any geologist's toolkit; all the more so in Great Britain where there is little exposure of the underlying rock over large parts of the country. Commonly, the geologist must infer what lies at depth from the surface expression of the lithology and structure through their effects on the topography. In the field, BGS geologists have done this for many years by employing a technique known as feature mapping. This consists of careful surveying of subtle breaks of slope, which can then be related to particular lithologies under the superficial cover. A classic example comes from the chalk of southern England (Fig. 1), where up to 12 distinct units can now be recognized by the effects that they have on the surface topography. Discontinuities in these surface features can commonly be related to faulting and other expressions of the underlying geological structure. These features are characterized by changes in elevation that can be seen in digital elevation data just as they can in the field. The data may be generated by digitizing contours acquired during topographical survey: from aerial photography, either as

a photogrammetric source for these contours or via the photogrammetric generation of a digital elevation model; or from direct measurement techniques, such as spaceborne, airborne or terrestrial radar or LiDAR (Light Detection And Ranging) sensors. Table 1 details the various elevation data used in the UK by the BGS over the past decade.

Depending on the technique used and the desired application, such data may represent the bare earth (a digital terrain model) or include the elevation of surface features such as vegetation (a digital elevation model or digital surface model). Geologists usually prefer to analyse the former, although the latter might give a more realistic visualization of landscape when used in combination with aerial photography.

Geologists increasingly use elevation data to help them accelerate feature mapping. This has several advantages, including the fact that it can be carried out at the desktop and then checked in the field. These data are often GIS-ready, and their interpretation can be carried out within standard GIS and digital mapping packages. The sun angle and topographical exaggeration can also be adjusted to emphasize subtle features. Several

Fig. 1. Subtle breaks in slope indicate the presence of different units in the Upper Chalk that can be seen in the laboratory using aerial photography and digital elevation data.

Table 1. *Strengths of main types of surface model used in the BGS, highlighting corporate utility of NEXTMap*

Elevation model source	Coverage	Z accuracy
Airborne LiDAR (Light Detection And Ranging)	Local	Centimetres
Aerial Photography & Digital Photogrammetry	Local	Centimetres–metres
NEXTMap Britain (using airborne IfSAR)	National	1 m
Ordnance Survey Landform Profile (5 m contours)	National	5 m
Centre for Ecology & Hydrology DTM (OS Panorama, enhanced so that rivers flow correctly)	National	10 m
Ordnance Survey Landform Panorama (10 m contours)	National	10 m

advanced image analysis techniques have been developed within the BGS that can, at least partially, automate the process of interpretation (Fig. 2). Elevation data are also used as the base on which other datasets, such as aerial photography, are visualized in their landscape context as part of the digital mapping process. The BGS has purchased national coverage of the NEXTMap® digital surface model dataset from Intermap™ as a BGS baseline dataset.

Terrain and elevation models. Geologists are most often interested in terrain, or bare earth, models, because these allow them to analyse the shape of the Earth's surface rather than those things growing or built upon it. A component of the NEXTMap® Britain dataset includes a bare earth model, but this has been generated from the original elevation model that was measured during the survey. This process involves editing the elevation data to remove features such as trees and buildings. Doing

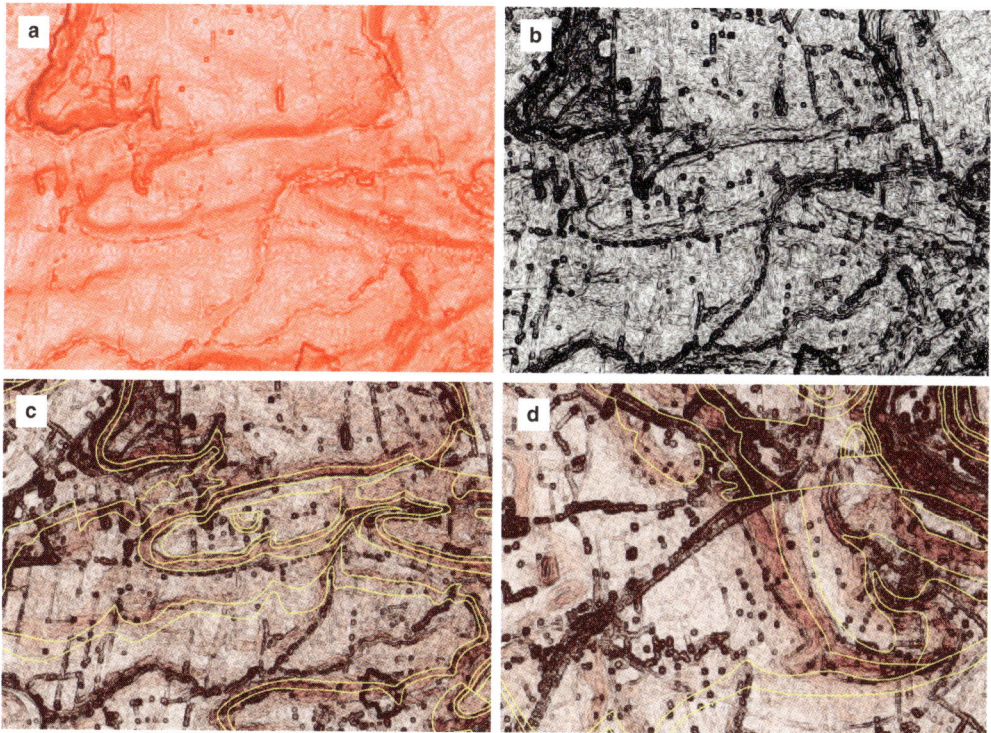

Fig. 2. Algorithms have been developed that find and emphasize breaks of slope; this information can help in assessing the accuracy of existing geological line work: (**a**) classify the DSM based on the slope angle; (**b**) calculate the rate of slope change; (**c**) combine and overlay the line work; and (**d**) reveal where the line work is in error.

this for the entire country requires automation and this is a considerable technical challenge. Consequently, the resulting bare earth model contains residual artefacts related to the features removed. Figure 3 shows a comparison between the terrain and elevation models to illustrate this. Small forest stands can confuse subsequent analysis techniques such as slope angle, giving apparent steep slopes at their margins. Such errors propagate through into derived products such as landslide hazard maps, giving erroneously high hazard values around forests. The BGS has attempted further editing of the NEXTMap® Britain dataset, using satellite imagery and photography to map the spatial distribution of vegetation on a national scale and highlighting potential problem areas for further editing. This artefact editing problem also affects elevation data extracted from satellite radar interferometry, which works on a similar basis to that from an aircraft, and stereo aerial photography and satellite imagery (e.g. stereo ASTER, SPOT or IKONOS data).

There are two other ways to tackle the problem. The first is to use nationally available contour data. As part of their topographical surveying process,

national mapping agencies commonly use digital photogrammetry to generate contours from stereo aerial photography. This involves the creation of a stereo model that contains an elevation model from the photography, from which contours can be digitized by a skilled analyst who can place the cursor onto the ground, even amongst tree stands. Unfortunately, it has been common practice not to extract and store the elevation model once the contours have been generated, otherwise this would be a valuable source of national elevation data. Instead, the contours can be used to work back to the elevation model by gridding and interpolation. As the contour data have already been generated in a way that avoids the recording of unwanted surface features, such datasets circumvent the artefact problem efficiently. However, such models tend to lack topographical detail owing to the degree of interpolation employed, especially in relatively flat ground where contours can be both poorly constrained and sparse.

The second approach is to use a data acquisition technique that measures the ground surface directly, penetrating the tree canopy. Airborne LiDAR data

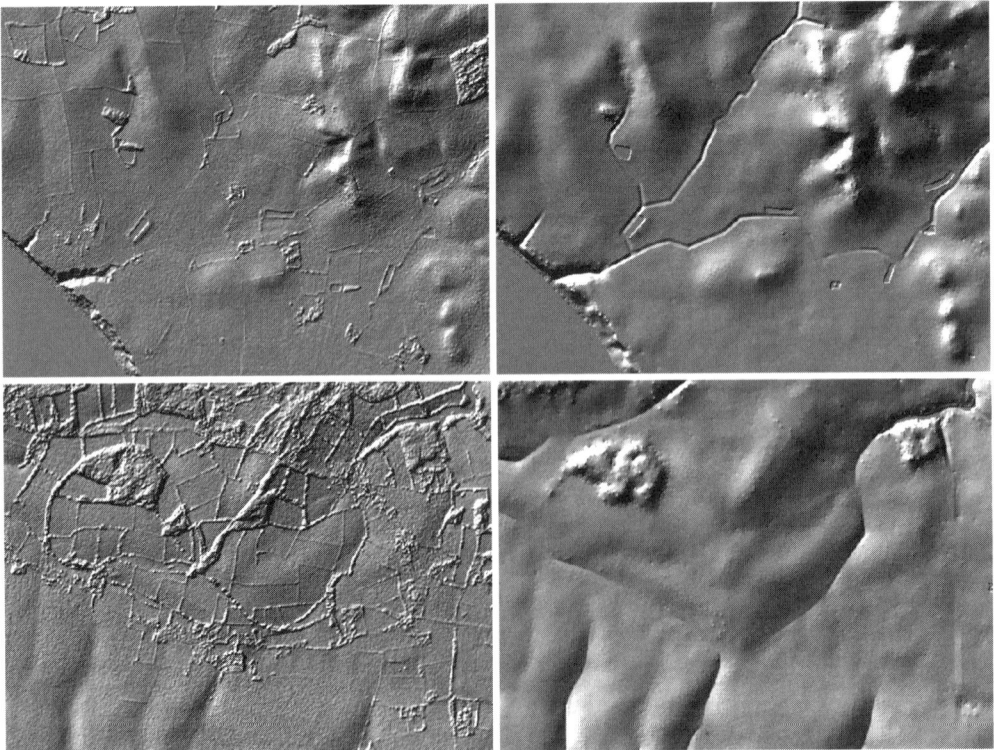

Fig. 3. Comparison of NEXTMap DSM (left hand images) and DTM (right hand images) for two areas on the Isle of Wight. The appearance of a river artefact can be seen in the top two images and forest stand artefacts are illustrated in the bottom pair of images.

have been used effectively for this purpose, even in rainforest areas. The laser in the aircraft is pulsed, so that every point on the ground has multiple measurements, at least some of which penetrate through the canopy and reflect off the land surface. Using such data, it is possible to extract the bare earth model from the last return, the structure of the tree canopy from the intermediate returns and the top of canopy elevation model from the first return within the same dataset. LiDAR data are also high resolution in x, y and z, commonly in the range of centimetres rather than metres, and they make a highly suitable dataset for geological terrain analysis. Their only drawback is that they are not yet available on a national basis; as ad hoc acquisition continues, the coverage in the UK is increasing to the point where national coverage has become possible to contemplate. Initiatives to pursue this are under discussion and it is likely to occur in the near future.

Imagery

Aerial photography has long been used by geologists to visualize the landscape in three dimensions, both in the laboratory using stereoscopes and in the field using field pocket stereoscopes or table stereo glasses. Before the advent of elevation data, this was the main way in which the topography was visualized so that desk-top feature mapping could be undertaken. In addition, imagery gives useful clues about lithology and soil type through the colour and texture that can be seen and associated with particular rock types. The patterns made by streams also vary depending on the lithology, and these patterns, together with textures, provide clues about jointing and fracturing. Major topographical lineaments that persist over kilometres are commonly associated with significant faulting or other geological structures.

The advances in computer processing power and storage in recent years have revolutionized the use of such data in geological mapping, and digital imagery is now a key dataset in the digital mapping workflow. The change started with the advent of satellite imagery in the 1970s and Landsat became an important reconnaissance tool for geologists, which still has its place today, particularly in poorly mapped, well-exposed terrain (Fig. 4). But in the UK it was the widespread availability of digital stereo aerial photography that began to see imagery take its place in the digital workflow. The geological survey has invested in national coverage from UK Perspectives and Getmapping survey companies, including monoscopic orthophotography for draping on elevation data and full stereo photography for more advanced analysis. This aerial photography underpins many mapping projects. Geologists interpret the landscape before going in the field, and then

use the field time for checking and investigating challenging or interesting areas. This generates both an economic and scientific return on the investment in the data. The national Landsat and aerial photography coverage form two further BGS Baseline Datasets.

Topography

Topography is the essential spatial backdrop to GSOs outputs. Geological maps are ultimately of little use unless they relate to an underpinning topography. It is also important in data acquisition as the spatial framework provided by topographical mapping references individual observations. The relationship between geological mapping and the topographical survey is so fundamental that the founder of the Geological Survey in Great Britain, Sir Henry De la Beche, was originally funded in 1832 by the Board of the Ordnance 'to cover the cost of geologically colouring the topographical maps of the Trigonometrical Survey'.

Modern topography. In Great Britain (GB) the principal supplier of modern topographical mapping is the Ordnance Survey (OS). The OS produces a wide range of mapping outputs at a range of scales down to 1:1250. The most important scales for most BGS geological mapping are 1:50 000 and 1:10 000. These scales are readily available in raster format, which is ideal for backdrops and locating observations using global positioning systems. The MasterMap product provides mapping at scales up to 1:1250. It is used occasionally for specific tasks but the cost of national coverage prevents its routine use by the BGS.

Historical topographical mapping. In Great Britain the Ordnance Survey was founded in 1791 and it has been publishing maps since that date. By the 1840s systematic surveying had commenced at the 1:10 560 scale for Great Britain. Many areas have been resurveyed repeatedly, producing a series of editions showing the changing landscape. This serial snapshot of the landscape of Great Britain provides valuable information to geoscientists. Information about the location of mine entrances, quarries and other excavations can be derived from historical topographical mapping. They also provide valuable information about anthropogenic landscaping. For example, consider a late-nineteenth century mine that produced a waste heap that was subsequently modified after closure of the mine, with associated tree planting to stabilize the slopes. Planners of subsequent developments need to be aware that the visible wooded hill is, in fact, composed of mine waste, with potential slope stability issues and the possibility that it contains toxic minerals or high-acidity materials and drainage.

Fig. 4. Landsat Mosaic of United Arab Emirates used to for geological mapping. Bands 741 in RGB.

Management and delivery of datasets

Once the datasets have been acquired they need to be managed rigorously within a controlled environment. The principal issues are:

- licensing and intellectual property rights;
- long-term storage and digital preservation;
- dataset limitations;
- communication.

Licensing and intellectual property rights

Acquired datasets commonly come with complex licensing agreements. It is essential that the terms and conditions of licences are understood and communicated to data users in a way that they can understand. Internal monitoring systems are required to ensure that research outputs, derived datasets and information products do not infringe third-party intellectual property rights. In the most difficult cases licensing agreements for a given product vary over time as the policies of the supplier change. This means that a dataset used in the creation of a specific research output might no longer be available for its continued use and exploitation, so the research output must be withdrawn from use. The BGS has found it simplest to attempt to acquire in-perpetuity licences for a given dataset in exchange for a one-off payment. This simplifies licence management and means that research outputs are more long-lived and robust.

Long-term storage and digital preservation

National digital datasets can be large, comprising multi-terabytes (terabyte $= 10^{12}$ bytes) of data. When such data are being acquired it is essential that the related storage issues are considered during the acquisition process. Does the organization have the storage and computer power to manipulate the datasets? Digital data formats change over relatively short timescales. Plans need to be in place to ensure that the dataset continues to be available, even if the original delivery format becomes obsolete. This is not simply a process of progressively migrating datasets to the current appropriate file format; the organization must have a clear understanding of any information losses that might take place during progressive file format changes.

Dataset limitations

No dataset is perfect. Each has its own limitations. Nevertheless, the temptation is to assume that digital datasets are perfect. Tarter (1992) has noted that '(the) myth of machine infallibility seems to create a demand for higher standards of quality for machine readable data than for traditionally distributed information'. Similarly, Peritz (1986) has suggested that 'the presumption of trustworthiness (of digital data) simply carries too much weight'. The reality is that data are not perfect, and dataset limitations need to be understood and documented. The aim of the documentation is to ensure that a potential user can assess whether a given dataset is fit for its intended purpose. For example, there is a predisposition to assume that digital raster images of historical Ordnance Survey 1:10 560 topographical maps have a similar accuracy to their modern 1:10 000 counterparts. However, there are considerable differences.

- The original 1:10 560 maps were paper prints and they are up to 150 years old. Unless stored in a perfect records management environment for their entire life, such paper maps can become distorted to varying degrees over time. The maps might, therefore, be spatially inaccurate before scanning.
- Different generations of maps will have been surveyed using different methods and/or instruments. The same geographical object will not necessarily be in the same spatial location on subsequent editions.
- The Ordnance Survey 1:10 560 maps are maps, not plans. They include cartographic generalizations that affect the spatial representation and location of geographical objects.
- Geographical objects change over 100 years. Buildings are extended or demolished and

rebuilt, changing their footprint in the process. Bridges and roads may be widened.

The cumulative result is that, potentially, a location determined from an historical map might be tens of metres from its correct location.

Other dataset limitations arise from the nature of the instrument collecting the data and the platform upon which the instrument was mounted. Where the platform is an aircraft and the instrument is IfSAR, a number of 'IfSAR artefacts' may be identified in the final dataset. Several known artefacts can persist despite processing of the data during and after acquisition. One type of artefact is 'motion ripples', which are caused by atmospheric turbulence preventing the aircraft from maintaining level flight during data acquisition. They appear as height ripples in the elevation data and as dark bands in the imagery at right angles to the direction of the aircraft's flight path. Processing eliminates most motion ripples, but some might persist into the final dataset (Intermap Technologies 2007).

Communication

The principal way to communicate information about a dataset is through its associated metadata. A rich, well-maintained metadata entry could enhance user understanding of the dataset. It is the dataset custodian's responsibility to develop and maintain metadata to meet the needs of users seeking to re-use and re-purpose the data. The profile and significance of metadata have risen in many countries in recent years following the introduction of new laws, including data protection and freedom of information legislation. Within the European Union (EU), directives have been issued relating to the use of public sector information and a common spatial infrastructure. These are being transposed into national laws by EU member states. All this legislation requires or implies that accurate, well-maintained, metadata are in place.

The sum is greater than the parts

One of the underlying motivations for the selection of the BGS national baseline datasets was the synergy between them. Elevation data can be analysed in their own right, but come to life when they are used as the backdrop for aerial imagery. Such imagery provides a unique view of the Earth's surface from above, but is far easier to interpret when draped over an elevation dataset to create a virtual, immersive environment than it is when using traditional stereo analysis techniques. This synergy extends the utility of the data from the specialist analyst to geologists (and, indeed, other scientists, professionals and the public) in general. Other datasets can be visualized far more clearly when viewed in their landscape context within a virtual environment.

Another synergy between elevation data and imagery or other raster datasets involves the generation of synthetic stereo imagery from ortho-corrected mono imagery. This approach allows other digital geoscience datasets, such as the BGS national digital geology and geochemical or geophysical datasets, to be viewed in stereo. Viewing geological map data in a new perspective against a 3D topographical model highlights any inconsistencies within the conceptual 3D geological models that underlay the original mapping, and allows for rapid correction or the targeting of field surveys to update maps in problem areas.

Elevation data have many applications in their own right, but really come into play as an underpinning dataset supporting the processing, display, interrogation and analysis of other geoscience data. This can include: their use to orthorectify other remotely sensed imagery; as a base on which to display 2D geological maps in 2.5 or 3D; or more complex algorithms within a model or GIS that take elevation, or a derivative like slope, as one input. The integration of elevation data into other geoscience workflows delivers substantial synergy. Satellite imagery can be placed in their real-world position and features extracted that can go straight into a GIS. Geological lines drawn in a predigital era can be seen in 3D and obvious errors corrected, releasing the potential of older datasets. Complex problems that may require a wide range of input parameters often have elevation, or a derivative, as a common thread. The latter point was illustrated during the first half of this decade when, one after the other, a series of Integrated Global Observing Strategies identified improved global DEMs as a high priority. They covered not only the geosciences (Marsh 2004) but also coastal observations, the water cycle, land observations and the cryosphere. Elevation data are one dataset that can pay dividends right across the environmental sciences.

Application

The applications of elevation data to the geosciences are many and varied. In the foundations of our science, the basic geological mapping requires the topography to be mapped first because the geology exerts control on the overlying landscape and an understanding of the latter helps reveal the former. In countries like the UK, large areas must be mapped without seeing a single exposure and the established technique, feature mapping, is essentially a detailed topographical analysis. Digital

topographical data have transformed this process in the past decade, allowing much of the critical information to be captured in the office. Fieldwork then focuses on the challenges and areas of particular geological interest. Beyond this, elevation data have an application in many other geoscience disciplines, at the least as a backdrop for other data. An important application area is geohazards, where slope and aspect are one of the key controls on ground instability. In pollution studies, the source–pathway–receptor model relies on topography to help determine the pathway and likely area for receptors. Applications exist in minerals geoscience, especially aggregate, resources and groundwater management. In fact, it is hard to think of a geoscience discipline where these data do not apply.

Case study: virtual field reconnaissance

GeoVisionary. BGS geoscientists have routinely used digital elevation models as part of their work for many years, but accessing and visualizing the data was often time-consuming and restricted by technology limitations to either small areas or low-resolution representations. In order to make full use of the BGS's new high-resolution baseline datasets (principally NEXTMap® Britain 5 m digital terrain model (DTM) and digital surface model (DSM) from Intermap™ Technologies, and aerial photography from UKP/Getmapping) a project was started in late 2006 – Virtual Field Reconnaissance – that aimed to create an environment in which to visualize and interact with all of these data. The project quickly built on BGS links with Virtalis Ltd, a British-based virtual reality company that had previously been commissioned to install an immersive 3-Dimensional Visualization Facility at two BGS sites and to produce custom geological visualization software.

The result of this collaboration is GeoVisionary, a software system that has built-in seamless streaming of multi-resolution levels of data, merging data such as existing digital geological maps, aerial photography, satellite imagery, field-slips, historical topographical maps, and subsurface 3D models, cross-sections and boreholes. The system allows teams of geologists to survey an area before commencing fieldwork, building an understanding of the terrain, which leads to a better interpretation of the geological structure. This initial assessment allows surveyors to effectively target fieldwork in areas where surveying is most required. On completion of fieldwork, surveyors can check their field interpretation in the virtual landscape. This team approach allows colleagues to work together on both pre- and post-fieldwork studies, better enabling communication, so increasing operational efficiency and enhancing scientific understanding.

In addition, whilst a variety of data can be visualized in the system, the elevation data are fundamental to its successful operation; they provide geoscience information in their own right but are also the back-cloth against which other data are displayed, and the top surface from which subsurface geological models are generated and hung.

Conclusion

The increasing variety and improved availability of national digital datasets are helping to provide exciting new tools for geoscientists. These data products, used in combination with innovative software, provide new ways for geoscientists to perceive and interpret the landscape. However, for the potential benefits to be achieved, a range of dataset management issues must be addressed. These include:

- licensing and intellectual property rights management;
- management of the digital datasets;
- digital preservation of the dataset;
- understanding the limitations of the dataset;
- communication of the above to geoscientists.

It is incumbent upon GSOs to make sure that they create and maintain efficient and well-supported information management systems to deal with these issues.

References

INTERMAP TECHNOLOGIES 2007. *Product Handbook & Quick Start Guide Standard Edition 4.2.* Intermap Technologies, Denver, CO.

JOHNSON, C. C., BREWARD, N., ANDER, E. L. & AULT, L. 2005. GBASE: baseline geochemical mapping of Great Britain and Northern Ireland. *Geochemistry: Exploration, Environment, Analysis,* **5**, 347–357.

MARSH, S. H. (ed.) 2004. *The Integrated Global Observing Strategy Geohazards Theme Report.* European Space Agency, Paris.

PERITZ, R. 1986. Computer data and reliability. *North west University Law Review,* **80**, 960.

TARTER, B. 1992. Information liability: new interpretations for the electronic age. *Computer/Law Journal,* **XI**, 484.

Index

Page numbers in *italic* denote figures. Page numbers in **bold** denote tables.